Lavoisier in
the Year One

GREAT DISCOVERIES

MADISON SMARTT BELL

Lavoisier in the Year One

The Birth of a New Science in an Age of Revolution

ATLAS BOOKS

W. W. NORTON & COMPANY

NEW YORK · LONDON

Copyright © 2005 by Madison Smartt Bell

For information about permission to reproduce selections from this book, write to
Permissions, W. W. Norton & Company, Inc., 500 Fifth Avenue, New York, NY 10110

Manufacturing by R. R. Donnelley, Harrisonburg
Book design by Chris Welch
Production manager: Julia Druskin

Library of Congress Cataloging-in-Publication Data
Bell, Madison Smartt.
Lavoisier in the year one : the birth of a new science in an age of revolution /
Madison Smartt Bell.— 1st ed.
p. cm. — (Great discoveries)
Includes bibliographical references and index.
ISBN 0-393-05155-2
1. Lavoisier, Antoine Laurent, 1743–1794. 2. Chemists—France—Biography.
3. Chemistry—France—History—18th century. 4. Chemistry—Nomenclature.
5. Chemical processes. I. Title. II. Series.
QD22.L4B25 2005
540'.92—dc22

2005002822

ISBN-13: 978-0-393-32854-7 pbk.
ISBN-10: 0-393-32854-6 pbk.

W. W. Norton & Company, Inc., 500 Fifth Avenue, New York, N.Y. 10110
www.wwnorton.com

W. W. Norton & Company Ltd., Castle House, 75/76 Wells Street, London W1T 3QT

1 2 3 4 5 6 7 8 9 0

To the spirit of
Giordano Bruno
Wherever it may be found

Contents

Acknowledgments

My thanks to James Atlas for the invitation; to Jesse Cohen for extraordinary editorial services; to Marie-Josèphe Mine of the Archives of the Academy of Sciences in Paris, for her help with the Lavoisier papers there; to Professor Esther Gibbs of the Goucher College Chemistry Department, for her help with certain facts of modern chemistry; and to Dr. James Hughes, for pointing me toward Condorcet.

Lavoisier in the Year One

I

Ancien régime

In early autumn of 1793, officials of the French National Convention called on Antoine Lavoisier at his private residence in Paris: 243 boulevard de la Madeleine. A block to the west along this fashionable street, work had been suspended, since 1791, on the church Saint-Marie Madeleine, designed in the style of the Greek Parthenon. From the unfinished classical portico, the view south was open down the rue Royale to the Place de la Révolution, where the guillotine would soon be installed. Dr. Guillotin, like Lavoisier a member of the Parisian scientific elite, had meant the device to be a humane and civilizing reform of the brutality of axes and ropes; he would be as appalled as anyone to see his idea evolve into one of the most horrifying weapons of terror that the Western world has ever known.

The officials had come, on behalf of the dread "Committee of Public Safety," to search and seize Lavoisier's papers; in the end they found nothing suspect but some correspondence in foreign languages (English and Italian) from fellow scientists: Lazzaro Spallanzani, Joseph Priestley, Joseph Black, Benjamin

Franklin. Lavoisier asked and was granted permission to apply his personal seal to the seized bundle. Probably he feared that more dangerous documents might otherwise be planted in the package by his enemies, though the report filed on the procedure stated that "it is not from distrust that he solicits this precaution, but for the sake of order."

The date, according to the French Revolutionary Calendar, was 24 fructidor of the Year One, though neither Lavoisier nor anyone else knew it. So far as they were all concerned, it was September 10, 1793. The French Revolutionary Calendar, though dated from the establishment of the French Republic on September 22, 1792, was not proclaimed and adopted until October of 1793. Therefore, the Year One existed only in retrospect; no one experienced it directly. And yet it mattered. Before the Year One, Lavoisier's life and career had been tidily woven into the social fabric of the Bourbon monarchy. From this day forward his life and work would be reexamined in the new and dangerous context of revolution and terror.

The French Revolutionary Calendar was intended to detach the authoritarian hand of religion from the measurement of time—and to purge the Julian calendar of mathematical eccentricity. The reformed calendar operated on base ten, dividing each month into three ten-day cycles, and each day into ten periods of one hundred minutes worth one hundred seconds apiece. Lavoisier himself was an advocate of the reformed calendar, and in 1793 he was actively at work on a parallel reform of the system of weights and measures; this decimalized method of measurement proved much more durable than the Revolutionary Calendar and remains standard, as the metric system, today. On 24 fructidor of the Year One, another of Lavoisier's scientific colleagues, Antoine-

François de Fourcroy, came with the other officials to boulevard de la Madeleine to repossess instruments Lavoisier had
been using for the weights and measures project—a bad
omen, for Lavoisier had calculated that his participation in
this public project would ensure his safe passage through the
turbulent times.

If Lavoisier had properly understood the danger that the
radical new social context of the as-yet-undeclared Year One
presented him, he would perhaps have fled the country.
Apparently he failed to perceive the extent of his risk,
although in other contexts he understood—better than any
other scientist of his time—the crucial importance of radical
changes in point of view. What Lavoisier called *le principe oxygine*, whose discovery would ensure his permanent and
prominent place in the history of science, had been discovered
by others before him: Joseph Priestley, who thought of it as
"fixed air," and Carl Wilhelm Scheele, who called it "fire air."
Lavoisier's radical act was to understand and define the not-
altogether-new gas as *oxygen*, thus placing it in an entirely new
context from which the whole of modern chemistry would
begin to evolve. His was an extraordinary exercise in the
power of naming.

IN THE YEARS before the French Revolution, Lavoisier's
genius for order had found many applications outside his
purely scientific research. In 1788, he held five important public posts at once, including the directorship of the Gunpowder
and Saltpeter Administration and a place on the board of the
Discount Bank, which gave him an influential role at the center of French national finance. By 1790, his omnipresence in
public affairs, along with his very considerable wealth, made

him an attractive target for the radical left. In January 1791, the Jacobin journalist Jean-Paul Marat assailed him in the pages of *L'ami du peuple*: "I denounce to you the Coryphaeus of the charlatans, Master Lavoisier, son of a land-grabber, apprentice-chemist, pupil of the Genevan stock-jobber Necker, a Farmer General, Commissioner for Gunpowder and Saltpeter, director of the Discount Bank, secretary to the King, member of the Academy of Science, intimate of Vauvilliers, unfaithful administrator of the Paris Food Commission, and the greatest schemer of our times."

Marat's denunciation sprang from old resentments and rivalries; in 1779 Lavoisier had helped discredit Marat as a scientific charlatan—on behalf of the French Academy of Sciences. In the leveling climate of 1791, the very existence of national academies for the sciences, literature, and culture could be made to seem culpably elitist, and the patent of nobility that Lavoisier's father had purchased for him as a wedding gift in 1771 had very much ceased to be an asset. But his fatal point of vulnerability, the one that brought the commissioners to his house on boulevard de la Madeleine, was what had been his most lucrative day job—working for *la Ferme Générale*, or the General Farm.

For centuries, the collection of French taxes had been leased or "farmed out" by the French monarchy to private investors, who guaranteed the Royal Treasury fixed sums for the period of each lease, then took their own profit or loss from the taxes they could collect. By the end of the seventeenth century, the "tax farm" had swollen into a behemoth with thirty thousand employees. Late in the eighteenth century, the French government had become deeply dependent on the General Farm to obtain credit and service a rapidly growing national debt.

The General Farm was managed by a corporation of between forty and sixty partners when Lavoisier bought into it in 1768. The price of a full partnership that year was 1,560,000 livres; at the age of twenty-four, Lavoisier bought a one-third share from the elderly tax-farmer Baudon, who was seeking an assistant, for a down payment of 68,000 livres. He approached the business of tax collection with the reformer's zeal he brought to almost all his activities, but even his most enlightened innovations tended to put new and unwelcome pressures on French taxpayers. The General Farm was no more popular than any tax authority in any other time and place, and probably less popular than most. The organization collected a tax on salt (*gabelle*) and another on alcohol and tobacco (*aide*), along with customs duties (*traites*) and duties on goods entering Paris from elsewhere in France (*entrées*). Evasion of all of these taxes through smuggling and other fraud was epidemic, and harsh punishment for such offenses increased the general distaste for the Farm. Moreover, accusations of profiteering were well founded.

The General Farm was abolished, amid widespread charges of mismanagement, in 1791. Lavoisier had resigned his position not long before; however, there was a demand in the French National Convention for an investigation of the Farm's affairs reaching all the way back to 1740. The Farm's assets were supposed to be liquidated and added to the national treasury, but this procedure kept getting pushed to the back burner by a series of political crises, as the Farmers, or ex-Farmers, were accused of stalling. By the fall of 1793, as the

Reign of Terror commenced, impatience to resolve the matter of the General Farm (and collect the proceeds) had become extreme. Lavoisier was only one among many erstwhile Farmers to have his papers searched and seized.

Participation in the General Farm made Lavoisier one of the most prosperous members of the bourgeois class that grew and thrived during the last two decades of Bourbon rule. By 1786 he had turned a total profit of 1.2 million livres—the equivalent of 48 million in twentieth-century U.S. dollars. His manner of living was not ostentatious for a man of such wealth; a financial declaration for 1791 lists six household servants (one cook, one chambermaid, one coachman, and three lackeys)—a fairly small staff, considering the time and his position, though he also owned a 1,400-acre country estate, Fréchines, in the Loire Valley, and another 254 acres at Villefrancoeur.

Lavoisier's career in the General Farm furnished him an excellent income, while leaving much of his time free for science and other pursuits. Taxes, in fact, supported his research—an abnormality for the period. In eighteenth-century France, science might be a vocation, but it was not much of a livelihood. Public financial support for science was scant; aspiring scientists had to bear the cost of their own research programs. Lavoisier, whose family background had been relatively modest, used his profits from the General Farm to equip one of the most sophisticated—and expensive—laboratories in Europe.

ANTOINE LAVOISIER's great-great-great-grandfather was a courier riding postal relays for the Stables of the King. His great-great-grandfather became master of a relay station and hostelry in the market of Villers-Cotterêts, a town some fifty miles northeast of Paris. His great-grandfather, Nicolas Lavoisier, was a bailiff in the local court system and prospered well enough to own several houses in the town. Nicolas's son, the chemist's grandfather, became an attorney in the court and married the daughter of a well-to-do notary in the town of Pierrefonds. *Their* firstborn son, Jean-Antoine Lavoisier, was sent to Paris to study law. On the retirement of his mother's bachelor brother, Jean-Antoine inherited his uncle's place as a solicitor in the Parlement de Paris: the highest court of the *ancien régime*. It had taken the Lavoisier family over a century to complete this upwardly mobile trajectory to membership in the newly constituted professional class of *robins*— "men of the robe," or lawyers.

Along with his post, Jean-Antoine inherited his uncle's house in the Marais district of Paris, and a bequest of forty thousand livres. He found a good match in Emilie Punctis, the reputedly beautiful daughter of a well-connected bourgeois family that had made a modest fortune as butchers. Through careful investment in land, Jean-Antoine increased the family wealth—such were the grounds, however slim, for Marat's description of him as a "land-grabber."

The Lavoisiers' first child, Antoine, was born on August 26, 1742, in the house of Jean-Antoine. A daughter, Marie Marguerite Emilie, followed two years later. When Emilie Lavoisier died in 1748, the widowed Jean-Antoine moved with his children, aged five and three, into the Punctis household. There, the children were cared for by their unmarried aunt, Constance Punctis, until Marie Lavoisier died, at the age of fifteen. Antoine Lavoisier, who would eventually die childless, was to be the last of his line.

A childhood marked by these losses left young Lavoisier quiet and sober, preferring study to play. At the age of eleven he enrolled in the Collège des Quatre Nations, where his father had earlier been schooled. Informally known by the name of its founder, Cardinal Mazarin, the Collège Mazarin occupied a magnificent domed building, directly across the Seine from the Louvre; today, as the Institut de France, the structure houses Lavoisier's papers, with other archives of the Academy of Sciences. The family plan was for Lavoisier to follow his father into a career at law. At the Collège Mazarin, he tried his hand at literature, attempting a drama in the style of Rousseau's *La Nouvelle Héloise*. In 1760, the year his sister died, he won a second-place prize for an essay (unfortunately lost) on the question, "Whether rectitude of the heart is as necessary as precision of intelligence in the search for truth."

Lavoisier's first exposure to chemistry came from his school's instructor, Louis C. de La Planche, but a more important influence during those years came from l'Abbé Nicolas-Louis de Lacaille, an astronomer and mathematician whose observatory was on the grounds of the Collège Mazarin. Lacaille had taken the radical step of publishing his own algebra and geometry textbook in French, deeming it superior to the customary Latin. With the aid of Diderot and the other Encyclopédistes, eighteenth-century French was indeed crystalizing into the most lucid of the European languages, the ideal medium for works of pure reason. Lavoisier, who along with astronomy learned calculus and elements of Newtonian physics from Lacaille, did not miss this lesson. His liking for rational order in all things is grounded at this point; later on he would write, "I was accustomed to that rigor of reasoning which the mathematicians put into their works"—and especially the rigorous, step-by-step procedures of geometrical proof.

Lavoisier left the Collège Mazarin in 1761 and enrolled in the Paris Law School, yielding to his father's argument that the sciences were an admirable leisure activity but did not amount to a profession. He was dutiful in his study of law, but more passionate in the scientific education he pursued at the same time. During his law school years he studied mineralogy with Jean-Étienne Guettard, a geologist in the Academy of Sciences who, though a self-proclaimed misanthrope, was also a regular guest in the Punctis household. At the Jardin du Roi, he studied botany with another well-known academician, Bernard de Jussieu, and took chemistry courses from Guillaume-François Rouelle. If Lavoisier had any taste at all for the flesh-pots of Paris during those years, his double program of study left him no time for them; he was retiring to the point of reclusiveness, reputed to pretend illness from time to time to avoid social obligations. A friend of his father's, M. de Troncq, sent him a bowl of gruel with an ironic admonition: "regulate your studies, and believe that one more year on earth is worth more than a hundred in the memory of men."

In 1764 Lavoisier received his legal degree and was admitted to the Parlement de Paris; however, he would never practice law. At the tender age of twenty-one, he had already begun to plot a course toward membership in the Academy of Sciences. The Academy had been officially established a century before by Louis XIV's chief minister, Jean-Baptiste Colbert (who was also responsible for the consolidation of the General Farm into a single organization), in order to create a formal framework for the community of French scientists that had so far evolved organically. Supported by royal patronage, the Academy's mission was to pursue both pure and applied science—to seek the prestige of discovery along with the material gains of scientific practice. Internally, the Academy functioned as a meritocracy, rewarding and promoting those whose contributions to science were strongest. Externally, it had the authority and the responsibility to validate or discredit new scientific discoveries and theories that were proffered to the public; by the mid-eighteenth century it was the ultimate arbiter of scientific progress. Like the literary and cultural academies founded around the same time, the Academy of Sciences enjoyed royal protection and some support from the Royal Treasury, while also preserving enough autonomy to place it out of reach of national politics—an important advantage, analogous to what we understand as "academic freedom" today. In a eulogy of Colbert he drafted in 1771, Lavoisier described the academies as "little republics," noting that "their active power also overwhelms any opposition arising from ignorance, superstition, and barbarism."

In 1764 Lavoisier began work on a project for street lighting in Paris for a competition sponsored by the Academy. Ever willing to use himself as a scientific subject, Lavoisier sequestered himself in a blacked-out room for six weeks to carry out his study. Deemed more theoretical than practical, his essay was nevertheless accepted for publication by the Academy, and a gold medal was struck to honor his achievement. In 1765, in the role of "visiting scientist," Lavoisier presented a paper to the Academy titled "The Analysis of Gypsum" (the essential ingredient of plaster of Paris); the Academy reviewers praised his work for the "ingenious explication by which it reduces the phenomenon of the hardening of plaster to the simple laws of crystalization."

Apolitical as the Academy was meant to be, election (as the word implies) was not without a political dimension. The apprenticeships Lavoisier had served with Academicians like Guettard and Jussieu gave him a solid base of support; other backers were friends of his father. The Academy had a fixed number of members, and vacancies at the lower levels were typically created by promotion to the higher ranks. Despite recognition of his promise and a considerable lobbying effort, Lavoisier was not nominated for the vacancy that occurred in the chemistry section in 1766, and for which he competed against much older scientists with longer-established careers.

Following this setback, Lavoisier returned to a project creating a mineralogical atlas of France that he and Guettard had begun earlier, and spent most of the next two years doing fieldwork outside of Paris. Meanwhile his father, who had apparently accepted Lavoisier's determination to make the sciences the center of his working life, continued to manipulate whatever Academic threads passed through his hands, maintaining a climate of receptivity for his son's efforts. In the spring of 1768, Lavoisier returned to Paris with new papers to present, one on "techniques for determining the specific weights of liquids" and another on "the character of the waters" in regions he had visited for the mineralogical survey. The pairing offered a nice balance of pure science with a practical application to the issue of the national water supply.

In May of 1768 Lavoisier and Gabriel Jars were nominated for a vacancy that had opened in March of that year. Because of his longer years of service, Jars was named to the vacant post, but Lavoisier was also offered immediate membership in the chemistry section as a "supernumerary adjunct," on the understanding that he would be confirmed in the next vacancy that occurred. In fact, Lavoisier had received a few more votes than Jars; the solution was arbitrated by the king. To the Academicians who might have looked askance at Lavoisier's involvement in the socially dubious General Farm, a Monsieur Fontaine retorted, "That's just fine! The dinners he will serve us will be all the better!"

THE AVERAGE ANNUAL stipend for a member of the Academy of Sciences was two thousand livres, not to be sneezed at but also well short of what it took to sustain a middle-class manner of living in Paris—and as a new member, Lavoisier could not have expected to earn so much from his Academy post right away. With money bequeathed to him by his mother, he bought his first shares in the General Farm a matter of weeks before his election to the Academy of Sciences. This step freed him from any need to support himself by the practice of law. In theory, at least, Lavoisier could live (more than comfortably) on investment income while dedicating his working hours to science; in fact, his involvement with the General Farm became the occasion for a considerable amount of public service work, often overlapping with his scientific interests and research.

He began as a regional inspector for the Farm's Tobacco Commission, combining his inspections with his exploratory trips for the mineralogical atlas in 1769 and 1770. Lavoisier's mission was to fight a vigorous trade in contraband tobacco, mixed by retailers with tobacco properly taxed by the Farm. He reported his results to Jacques Paulze, a senior partner in the Farm who was also, like Lavoisier's father, a lawyer in Parlement de Paris.

In 1770, soon after the death of his wife, Paulze brought his daughter back from the Montbrison convent, where she had been educated, to his Parisian domicile. Marie-Anne Pierrette Paulze, thirteen years old, was meant to serve her father as hostess. Antoine Lavoisier, then twenty-seven, was thrown into her company. Marie-Anne was her father's only daughter, as Lavoisier was an only son, and like Lavoisier, she had lost her mother at an early age. Despite her youth, she played her domestic role with confidence and grace, and she was attractive, with "very blue eyes, brown hair, fresh coloring and a small mouth."

Eighteenth-century bourgeois marriages were business deals first and romances later—if at all. Marie-Anne Paulze, who was a significant heiress among her other advantages, became the target of a hostile takeover attempt. The suitor was the comte d'Amerval, an impoverished fifty-year-old nobleman with a name for dissipation. Marie-Anne herself described him as "a fool, an unfeeling rustic and an ogre." In a letter to his wife's uncle, the abbé de Terray, Jacques Paulze toned down Marie's reaction to "a decided aversion," and said, quite firmly, that he would not force her into a marriage she disliked. Attracted by Amerval's title and under the influence of Amerval's sister, the baroness de La Garde, Terray continued to press for the match. As controller-general of finances, Terray had the power to remove Paulze from the directorship of the General Farms department of tobacco, and threatened to do so. With the aid of other allies, Paulze protected his job, but he saw that his daughter would remain vulnerable so long as she remained unmarried.

In age, temperament, and financial position, Antoine Lavoisier was a much more appealing match for Marie-Anne than Amerval, and not at all displeasing to the girl. The two were married on December 16, 1771—the bride fourteen, the groom exactly twice her age. Gracious in his disappointment, Terray attended the ceremony as a witness for the bride, who had spent barely a year at home between the time she left the convent and her marriage.

Madame Marie-Anne Lavoisier grew to be a person of substantial talent and accomplishment. She seems to have shared, at least to some extent, Lavoisier's retiring disposition and his preference for productive work. Although she began a long-term affair with one of her husband's colleagues, Pierre-Samuel Dupont, in 1781 (after thirteen years of marriage), she conducted it with such discretion that no one seems to have suspected it until after her husband's death. And her devotion to Lavoisier's career was unfaltering. Childless, she served Lavoisier as a very efficient laboratory assistant, if not a manager. Lavoisier gave her her first lessons in chemistry, and later on she studied the subject with Jean-Baptiste Bucquet. She kept meticulous notes on experiments and helped draft numerous of Lavoisier's texts. Her knowledge of Latin and English enabled her to translate articles and, when necessary, entire books. Many illustrations of Lavoisier's equipment and materials came from her hand; her nicely composed sketches of his respiration experiments include her as a working member of the team.

Marie-Anne's considerable talent for drawing led her to become a pupil of the icon of French neoclassical painting, Jacques-Louis David. Apart from the scientific illustrations, little of her artwork has survived, other than a portrait of Benjamin Franklin which she presented to him as a gift in 1788. Though she appears not to have cared much for society for its own sake, during the twenty-three years of their marriage, she

Drawing by Madame Lavoisier of one of Lavoisier's experiments on respiration, involving a subject performing an activity. Madame Lavoisier is on the right, making notes.

made the Lavoisier household one of the most important scientific salons in Paris. Her role in Lavoisier's life was memorialized in a poem by Jean-François Ducis:

> *Épouse et cousine à la fois,*
> *Sûre d'aimer et de plaire,*
> *Pour Lavoisier, soumis à vos lois,*
> *Vous remplissez les deux emplois,*
> *Et de muse et de secrétaire.**

*Wife and cousin at the same time
Certain to love and to please
For Lavoisier, in thrall to your rule
You fill the two roles
Secretary and muse.

———

FROM EARLY DAYS Lavoisier had a strong interest in applications of science to the public good. His study of Parisian street lighting was followed in 1768 by an analysis, on behalf of the Academy of Sciences, of a proposal by Antoine de Parcieux for an aqueduct to bring water into Paris from the river Yvette—relieving citizens of dependence on the polluted waters of the Seine. Due to lack of funds, and to Parcieux's death, the aqueduct was never built, but the analysis of water quality gave Lavoisier his first concentrated look at the composition of water and bolstered his interest in sanitation issues. When the aqueduct project failed, he began to investigate methods for purifying the water of the Seine. He understood that clean air was as important to health as clean water, and ventilation was key to his study for a reconstruction of the Hôtel-Dieu hospital, burned in 1772, and a study for a hygienic reform of Parisian prisons he directed in 1780.

As Lavoisier was the enemy of disorder and dirt in matters of public health, so too was he the enemy of chaos and disorganization in matters of administration. His temperament led him to discover principles of order wherever they existed, and to impose them where they did not. As tobacco inspector for the General Farm, Lavoisier developed a chemical test for adulteration: ash mingled with tobacco would reveal itself by an effervescent reaction to acid solutions such as spirit of vitriol. He disliked the contamination of tobacco with ash not only as a form of tax fraud but also as a hazard to the public health, and at the same time he was sensitive to the severity of the punishments for such offenses. Adulteration detected at the point of sale might have taken place somewhere else along the supply line; to isolate it, Lavoisier imposed a meticulous system of weighing and accounting at every stage of tobacco's journey to the retail market. This method for determining exactly when and how a change took place was much the same in its principle as the method he was learning to use in chemistry.

His work for the General Farm gave Lavoisier a vested interest in precise accounting, an interest that carried over into abstract economics. Heavily invested as he was in the existing tax system, he was not blind to its flaws or to the way in which those flaws impeded growth of the French economy. In the eulogy of the finance minister Colbert that he began in 1771 for another Academy competition, Lavoisier seems to admire his subject most for an ability to bring order out of anarchy: "In the midst of the chaos that surrounded him, and sustained by his own courage and the profound truth of his perceptions, he conjoined the law to the good." That conjunction could serve as a definition of rational idealism, and probably did define Lavoisier's sense of his own mission as a social reformer in several spheres, including economics and finance. Rational idealism implies its own morals, and the sum of Lavoisier's actions seems to prove that in fact "*la droiture du coeur*"* was fully as important as "*la justesse de l'ésprit*"† in the quest for truth.

In his examination of Colbert, Lavoisier began to understand money as "a fluid whose movements necessarily result in equilibrium," borrowing this analogy from physics. The exact balancing of precisely measured quantities by which Lavoisier learned to illuminate the affairs of the General

* "rectitude of the heart"
† "accuracy of the mind"

Farm, and later used in the larger financial affairs of the French state, applied just as well to his work in chemistry. In the words of Charles Gillispie, a biographer, "to finance, munitions, and science alike he brought the luminous accuracy of mind that was his signet, the spirit of accountancy raised to genius."

IN 1774 ANNE-ROBERT-JACQUES TURGOT replaced the abbé de Terray as controller-general of finance. Lavoisier was assigned by the Company of General Farmers to prepare a report on the Farm that Turgot had requested. Turgot was influenced by the French economic school called the Physiocrats to regard economics as an exact science, with laws as definite and permanent as those recently declared by Newtonian physics. Economic policy could thus be understood as a scientific application. These views were thoroughly congenial to Lavoisier, whose purely scientific ambition was to furnish chemistry with the clear definition physics had attained not long before. He cooperated enthusiastically with Turgot's initiatives, though many other General Farmers felt threatened by them. In fact, the hectic pace of Turgot's reforms menaced so many established interests that Turgot was dismissed in 1776, after a mere twenty months as controller-general.

The Swiss-born banker Jacques Necker, whose essay on Colbert had won the Academy competition from which Lavoisier had withdrawn, succeeded Turgot. Necker's reputation as a ruthlessly cutthroat financier was not undeserved, but he seemed the man for an hour when "a Controller-General is expected principally to find money and do nothing more." And after Turgot's thrilling, brief tenure, conservatives must have been reassured by Necker's declaration that reforms should be undertaken "without too grand an ardor."

One of Turgot's last acts as controller-general was the creation of the *Caisse d'Escompte*, or Discount Bank—a private institution, but with the important function of lending money to the Royal Treasury. In 1788 Lavoisier became a director of the Discount Bank. By 1789, as the first phase of the French Revolution unfolded, national finances were in crisis, and the private stockholders of the Discount Bank grew unwilling to lend more money to a government whose debt to the bank was already heavy. Jacques Necker responded with a proposal to nationalize the bank. Lavoisier, by then president of the bank's board of directors, supported the plan, but it failed in the National Assembly. Indeed, because of the Discount Bank's tenacity in protecting the stake of its private investors, it came under vigorous attack from the increasingly radicalized Assembly, with a faction led by Mirabeau demanding its liquidation.

By then the Discount Bank was in such a shaky situation that it could no longer redeem its own banknotes for cash, but Lavoisier and his fellow directors argued that the bank had been forced into this predicament by the government's failure to repay loans, and that without the bank's support, the Royal Treasury would have collapsed and the whole French economy with it. It was a period of French food shortages verging on famine, requiring importation of wheat from abroad to make bread—the era of Marie-Antoinette's famously unfortunate remark that if there was no bread for the poor, "Let them eat cake." The real cake-eaters were just beginning to suspect the hazards the time held for them; the French state's desperation for money would soon become a dire threat to private wealth and the people who owned it.

In the event, the notion of nationalizing the Discount Bank was superseded by a proposal by Mirabeau: that the state begin to issue its own paper money. These notes, called *assignats*, were supposed to be secured by lands that were seized from the church as the first wave of the Revolution rolled over France in the summer of 1789. Though Lavoisier disapproved of Mirabeau's plan, he joined a committee to supervise the *assignats*, devoting most of his attention to the prevention of counterfeiting, with which the Discount Bank had also been plagued.

Though the General Farm had not yet been formally abolished, the taxes it collected had been repealed by then, and the rest of the tax system was in similar disarray. Licensing an impoverished government to print its own money had the predictable result. Repayment of its loans in *assignats* further undermined the paper issued by the Discount Bank. As both *assignats* and banknotes lost value, and the price of bread climbed absurdly high, public opinion turned further against the fiscal conservatives, Lavoisier among them, who had tried to stop the inflationary spiral from beginning in the first place. The National Assembly resolved to reclaim public finances from the hands of "the great corps of venal financiers."

In September of 1790, as the atmosphere darkened, Jacques Necker precipitously left France. As Marat's diatribe underlined, the collegial relationship Necker and Lavoisier had enjoyed now became a dangerous association to the man who stayed behind. Though Lavoisier's conduct in finance had been impeccable, the fact that he simultaneously held positions in the Discount Bank and the Royal Treasury, to which he had been appointed in April 1790, suddenly looked suspicious. Lavoisier decided to decline his salary from the Treasury, and on April 9, 1791, he made the dubious move of announcing his decision in an open letter to *Le Moniteur*. "The emoluments I enjoy as Administrator of Gunpowder," he wrote, "precisely because they are moderate, agree with my manner of living, my tastes and my needs, and at a moment when so many honest citizens are losing their security, I could not for anything in the world consent to profit from a double salary." This program of retrenchment was probably sound, but announcing it publicly was a misstep that only brought more hostile attention to the point that while Lavoisier's manner of living could fairly be described as "moderate," the extent of his wealth and resources could not.

Like his involvement in French public finances, Lavoisier's interest in the making of gunpowder originated in the overlap of his work for the General Farm with Turgot's short-lived ministry. In the 1770s the Farm charged him with inspections not only for the tobacco tax but also for the salt tax, and since saltpeter refiners liked to evade taxes on the sale of the salt, which was a by-product of their process, Lavoisier got involved. The royal prerogative to collect saltpeter, though not part of the General Farm's authority, had traditionally been leased out to private interests on a similarly antique and cumbersome basis. The French state had a right to search for saltpeter anywhere, and saltpeter collectors' license to invade private property (even to scrape saltpeter from the interior walls of private houses) was fraught with abuse.

Of course there was a national interest in the supply and quality of French gunpowder, which, before Lavoisier and Turgot turned to the matter, had become notoriously unreliable. Indeed, France had been obliged to bring the Seven Years' War to a hasty conclusion, on unfavorable terms, because, thanks to a saltpeter shortage, the gunpowder supply had run out. As an inspector for the salt tax, Lavoisier discovered the manifold inefficiencies of the method of gunpowder production then in use and reported them to Turgot, who rescinded the lease of the Gunpowder Farm and authorized Lavoisier, Henri François d'Ormesson, and Pierre-Samuel Dupont to create a new Régie des Poudres et Salpêtres (Gunpowder and Saltpeter Administration).

Lavoisier, who wielded executive authority for the new Gunpowder Administration, immediately suppressed the search and seizure rights of the saltpeter collectors. In 1775, with Turgot's support, he organized an Academy of Sciences competition for studies on both the natural formation and the artificial production of saltpeter. His mineralogical travels with Guettard gave Lavoisier some background in the former; concerning the latter, it was convenient that one of France's very few cultivations of artificial saltpeter was on the outskirts of his family's old hometown of Villers-Cotterêts.

Under Lavoisier's direction, gunpowder research became a military science project with strong government backing. His series of programs worked toward better methods of extracting natural saltpeter, more efficient techniques for producing artificial saltpeter, and a more precise understanding of the chemistry of good gunpowder. These programs were so successful that by 1776 France had surpluses of gunpowder to offer to the American revolutionaries. In 1789 Lavoisier could claim, with some justice, that "North America owes its independence to French gunpowder."

The closest gunpowder factory to Paris was nearby at

Essonne; there the new research and development programs were put to the test. The work was not without its hazards. In 1788 the chemist Claude-Louis Berthollet discovered that a more powerful gunpowder could be made using a volatile muriate of potash. Lavoisier took a group of visitors on a tour of the Essonne powderworks on a day when the new formula was being tried. Though he had instructed his party to remain behind a safety barrier, not all of them obeyed, and when the mixture exploded, one of the chemists and a woman visitor were killed. Work at Essonne continued nonetheless; Éleuthère Irénée Dupont (son of Pierre-Samuel) got his first training on gunpowder-making there, and later on founded the Dupont chemical dynasty in the United States with a powder mill in Delaware.

When Turgot lost his office in 1776, all his reforms were threatened. Lavoisier maneuvered hard, and successfully, to protect the Gunpowder Administration as a government monopoly. The national need for a reliable supply of quality gunpowder was so great that it was best administered by a centralized authority. Lavoisier argued in another unpublished essay that though "every exclusive privilege is without question a violation of the natural order," it was justified in this case by the urgency of the need and the success of the result.

It was not only the public good that attached Lavoisier to the Gunpowder Administration, which soon after its creation erected a new building on the grounds of the Paris Arsenal on the right bank of the Seine: the Hôtel des Poudres et Salpêtres. There Lavoisier built himself a state-of-the-art laboratory, which would be the theater of all his research and experimentation for the rest of his working life. For that reason, the

Gunpowder Administration became the one of his several public posts he was most determined to retain when the tide turned against him in 1791. But in 1776 this position was more easily defended and, in fact, became quite a comfortable one for both Lavoisier and his wife, who moved to a private apartment in the Petit Arsenal in April of that year.

This situation meant that for all practical purposes Lavoisier woke up in his laboratory every day. His daily routine was as orderly as the mind of the man. Rising at 5:00 A.M., he devoted three morning hours to pure science in the lab, then spent the hours from 9:00 A.M. to 7:00 P.M. on affairs of the General Farm, the Gunpowder Administration, and the Academy of Sciences, housed at that time in a capacious suite of rooms in the Louvre. Another three evening hours were devoted to scientific work in the Arsenal laboratory. In addition, Lavoisier spent all of his Saturdays there, with a growing coterie of students. Marie-Anne Lavoisier, who though not yet twenty was vigorously acquiring the broad set of skills that would make her indispensable to her husband's research programs, wrote of Lavoisier's scientific Saturdays, "It was for him a day of happiness; a few enlightened friends, a few young people proud to be admitted to the honor of participating in his experiments, gathered in the laboratory in the early morning; it was there that we lunched, there that we held forth, there that we created that theory which has immortalized its author."

These were productive and happy years for Lavoisier and his wife. Lavoisier's adolescent shyness seems to have diminished as he settled into his marriage and began to become a quite prominent public figure. The couple went out frequently to the Paris Opéra and, because of Marie-Anne's interest in painting (she had recently begun her lessons with David), to art exhibitions. Madame Lavoisier's skills as secretary, lab assistant, promoter, and publicist were matched by her social graces, and the Lavoisier salon at the Arsenal was frequented by the most illustrious members of the international scientific community: Joseph Priestley, Joseph Black, Martinus Van Marum, Horace de Saussure, and Benjamin Franklin, who was the toast of Paris in the years following the success of the American Revolution, and whom the Lavoisiers especially courted.

The Arsenal laboratory remained at the center of Lavoisier's life even after he removed his private domicile to boulevard de la Madeleine. He also remained one of those responsible for the large powder depot there. Some of the Arsenal buildings actually communicated with the adjacent fortress of the Bastille; powder for the Bastille was stored in the Petit Arsenal near the Lavoisier apartment (a circumstance that may have inspired Lavoisier to design new magazines that would discharge their contents harmlessly through the roof in case of accidental explosion). On the twelfth and thirteenth of July in 1789, the commander of the Bastille became concerned that the Arsenal magazines might be blown up or their contents stolen by revolutionary rioters, and ordered that the Arsenal powder be transferred to the fortress. Though Lavoisier and his fellow directors had no choice but to comply, their obedience brought them under suspicion as "enemies of the people" who had schemed to deprive the Paris populace of the powder they needed to overthrow the monarchy.

The Bastille fell on July 14, and by August, mob rule had a footing in Paris. At the first of the month, the Arsenal magazines were clogged with inferior powder being transshipped to

Rouen and Nantes for sale to slavers. Lavoisier was supervising the loading of a barge with powder cases marked "*poudre de traîte*." The word *traîte*, which means "trade," had become the commonly used term for the slave trade. This innocuous label was interpreted by the large, excited, half-literate crowd that gathered to mean "traitor's powder." A cry went round that the barge was charged with munitions to put down the Paris Revolution. Hysteria erupted and Lavoisier, with another Gunpowder director, Jean-Pierre Le Facheux, found himself captive of the mob. The frustration of this eminently rational man as he tried to explain the idiotic misreading of the "trade powder" label to his captors can only be imagined.

During the July gunpowder crisis, Lavoisier and his colleagues had held out a week's supply of powder for the National Guard, which now appeared to protect them from being hung from lampposts en route to the Hôtel de Ville. The crowd poured into the hall for a public debate in which Lavoiser finally managed to convince his audience of the harmlessness of his own actions. The crowd's ire then shifted to the marquis de La Salle, the National Guard commander who had signed the order for the gunpowder shipment, and in this fresh burst of confusion, Lavoisier managed to slip safely away.

It was a narrow escape, and a lucky one; summary mob executions had become frighteningly common in the summer of 1789. Lavoisier and his friends were discouraged from publishing a justification of their conduct, and such a shadow of suspicion lay over these wholly innocent events that the mere mention of the Gunpowder Administration in Marat's anti-Lavoisier diatribe had an opprobrious tinge.

———————

As a rising administrative star in the General Farm, Lavoisier was concerned not only with the taxes on salt and tobacco but also with the duties on goods entering Paris from outside the city. Paris had long since sprawled beyond its old barriers, and the absence of a clear system of checkpoints was complemented by a bewildering disarray of collection and inspection practices involving twelve hundred confused employees: another chaos for Lavoisier to organize. Lavoisier had proposed the construction of a new wall around Paris as early as 1779, but the idea lay dormant until 1783, when he was appointed to the General Farm's central administrative committee, which had direct responsibility for tax collection at the Paris tollgates. The central committee supposed that 20 percent of goods coming into Paris slipped silently through tears in the tax net, for a loss of 6 million livres a year.

In 1787 the highly regarded architect Claude-Nicolas Ledoux was commissioned to design and build the Farmers' tax enclosure—an elaborate and very expensive project featuring a heavy stone wall six feet high, punctuated by sixty-six elegant Palladian pavilions serving as tollgates—each with a different design. Though fifty-eight of Ledoux's gates were completed, only four survive today: in the Parc Monceau, at the Place de la Nation, at the southern terminus of the bassin de la Villette, and above the catacombs in the Place Denfert-Rochereau.

The wall was horrendously expensive—at 30 million livres, it cost five times the annual revenue loss it was intended to prevent—and despite its architectural grandeur, it was vastly unpopular. Louis-Sébastien Mercier adapted the public sentiment into a sardonic quatrain:

Pour augmenter son numéraire
Et raccourcir notre horizon
La ferme a jugé nécessaire
*De nous mettre tous en prison.**

An anonymous pamphlet suggested that while the General Farm might want to raise a statue to Lavoisier on the ramparts of its wall, the Academy of Sciences should blush for the association. The same pamphleteer claimed that the duc de Nivernois, a marshal of France, said when asked what he thought of the new wall, "I am of the opinion that its author should be hanged." A multilayered pun of unknown authorship began to run through the Paris salons: "*Le mur murant Paris rend Paris murmurant.*"†

By the summer of 1789, Paris was muttering and grumbling over more than just the tax wall, which remained, however, an obvious and convenient target. The Farmers' wall was just short of completion on July 13, when (in the midst of the first gunpowder crisis at the Arsenal, and scarcely twenty-four hours before the sack of the Bastille) the Paris mobs began tearing it down, setting fire to most of Ledoux's lovely tollgates in the process. Two years later, the wall controversy helped fuel another attack on Lavoisier from Marat in his pamphlet *Modern Charlatans*: "If you ask what he has done to be so much extolled, I shall reply that he has procured himself

---

*To increase its gelt
and curtail our horizon
The Farm has judged it necessary
to put us all in prison.

† "the wall walling up Paris makes Paris grumble."

an income of a hundred thousand livres, that he has contributed the project to make Paris a vast prison, and that he has changed the term acid into oxygen, the term phlogiston into nitrogen, the term marine into muriatic, and the term nitrous into nitric and nitrac. Behold his titles to immortality. Proud of his lofty deeds, he now sleeps on his laurels."

MARAT'S ANIMOSITY TO Lavoisier dates back to 1779, when the firebrand journalist and provocateur (as history would finally remember him) made a serious effort to win the respect and celebrity in the sciences that Lavoisier had already begun to enjoy. In April of 1779, Marat demonstrated to Benjamin Franklin and the Academicians Balthazar Georges Sage, Jean-Baptiste LeRoy, and Trudaine de Montigny a series of optical experiments that purported to make visible the "matter of fire" or "igneous fluid." He had concocted a theory to explain the phenomena he displayed: "that fire resulted from the activation of particles of the igneous fluid contained in the bodies." The Academic observers praised the ingenuity of the experiments but declined to comment on the theory.

Marat continued to press for Academy recommendation of his theories about light, color, and fire. But when Lavoisier was appointed to a new academy team of observers, Marat objected to his presence. Lavoisier had himself begun a much more rigorous program of experiments in combustion in 1772, and Marat might have been right to anticipate that Lavoisier would not be impressed by his ideas and demonstrations. And in any case the demonstration that Lavoisier would have observed was postponed for lack of sun. Pressed by Marat for an endorsement of his experiments, the Academy of Sciences finally answered, on May 10: "It would be useless to go into detail to make them known; the commissioners do not regard them to be of a nature such that the Academy could give them its approbation or support."

One month later, an article in the *Journal de Paris* reported Marat's experiments and theories on the "matter of fire" as if the Academy had approved them. It was Lavoisier who noticed the false assertion and prompted a public repudiation. Definitively rebuffed by the Academy of Sciences, Marat developed a vivid hostility to the scientific establishment in general and Lavoisier in particular, and he took whatever opportunity he could find to attack both.

IN FACT, MARAT'S fantasies about light, color, and "igneous fluid" were not far out of line with other pseudo-scientific postulations of the day. Part of the mission of the Academy of Sciences—a project in which Lavoisier took special interest—was to arbitrate what was scientifically legitimate and what was not. Meanwhile, the whole definition of scientific orthodoxy was undergoing changes that Lavoisier himself described as revolutionary.

One of Lavoisier's first assignments for the Academy was an investigation of the divining rod—the tool of a persistent folk technique by which underground water is supposed to be detected by movements of a forked stick held by the diviner. Lavoisier's debunking of the practice was reasonably tactful: "There is water almost everywhere, and it is rare that one digs a well without finding some. So there is nothing out of the ordinary in the facts which are reported and to which one is tempted to attach a great importance; as this rod sometimes turns by involuntary movements on the part of the person who holds it, it is possible that some persons of good faith have been deceived and have attributed to an external cause an effect which depends only upon themselves."

Fifteen years later, in 1784, Lavoisier served on a committee that included among other scientific eminences Benjamin Franklin and the soon-to-be-notorious Dr. Guillotin, and that was charged with the investigation of "animal magnetism," a healing method made extremely fashionable in France by its leading practitioner, Anton Mesmer. In later years, "mesmerism" became synonymous with hypnotism, and it is worth noting that in our own time the effects of hypnotism, while understood to be psychological rather than physical, are accepted as being quite authentic and are used in therapies not greatly dissimilar to those Mesmer practiced in the eighteenth century. But if Mesmer knew he was practicing hypnotism, he did not say so. He claimed, and most likely believed, that his effects were produced by the manipulation of an invisible energy, like electricity, on which Franklin had done celebrated work, or the matter of fire, which turned out not to exist any more than Mesmer's animal magnetism did.

Mesmer's method of induction, which had been disseminated across an extensive network of practitioners by the time the Academy of Sciences took an interest, resembled a spiritualistic seance. Participants sat around a low tub full of damp sand, bottles of water, iron filings, rods, and other magnetic material. So as to be "magnetized," they grasped each other's thumbs and were also connected to each other and the iron rods by a cable. Soft music played while the mesmerist used flowing passes of his hands to manipulate the magnetic fluid that was supposed to suffuse this environment. Since most people have some susceptibility to hypnotism, most subjects of these procedures would have been likely to experience some level of hypnotic trance.

Lavoisier brought to bear on mesmerism the meticulous method of evidence sifting he used everywhere else. Certain effects of mesmerism, like the coincidental presence of water beneath the witch's wand, could not be denied (convulsive seizures could be reliably produced in the more susceptible subjects and were offered as evidence of "magnetism"). What stood to be challenged were the mesmerists' claims about causes. Through process of elimination, Lavoisier established that the same effects could be produced just as reliably without any magnets, simply through suggestion and touch; "imagination in the absence of magnetism produces all the effects attributed to magnetism," Lavoisier wrote; "magnetism without imagination produces no effects."

The committee reported that mesmerism was bunk, and on that basis the government launched a largely successful campaign to extirpate what amounted to a French mesmeric cult. Believers in mesmerism were hard to dissuade—the comparison to hypnotism suggests that mesmerism did produce real therapeutic results for many, though its explanations were spurious. The believers were numerous and some of them were powerful and would become more so during the revolutionary years. They included Jean-Paul Marat and his friend Jacques-Pierre Brissot, who like Marat had been frustrated in his efforts to penetrate the recently redefined scientific community and had turned to increasingly incendiary journalism. Brissot, who would occupy an extremely influential role in the Jacobin government before the Terror, began to target the Academy of Sciences (which did after all derive its authority from the throne) as a tyrannical institution. "The empire of the sciences must know no despots, no aristocrats, no electors," he wrote. "To admit a despot, or aristocrats, or electors officially empowered to set the seal on the production of genius, is to violate the nature of things and the liberty of the human mind."

It is more likely than not that Lavoisier approached mesmerism without preconceptions. He had trained himself to be resistant to received ideas. The invisible energy of electricity had recently been proved as an authentic physical phenomenon, and Lavoisier's own most crucial research program revolved around the investigation of the matter of fire. He took the investigation of mesmerism quite seriously, and a paper he wrote on the subject contains one of the most articulate statements he ever made about his own views of scientific rigor:

> The art of drawing conclusions from experiments and observations consists in evaluating probabilities and estimating if they are large and numerous enough to constitute proofs. This type of calculation is more complicated and more difficult than we think; it demands a great sagacity, and it is in general beyond the powers of ordinary men.
>
> It's on their errors in this sort of calculation that the success of charlatans, sorcerors, and alchemists is founded; likewise, in other times, that of magicians, enchanters, and in general all those who have deluded themselves or who seek to abuse public credulity.

There is more than a touch of aristocratic hauteur to be read between these lines, but perhaps Lavoisier himself did not notice it.

II

Out of Alchemy

Lavoisier first flirted with chemistry in his teens at the Collège Mazarin, where Louis de La Planche taught a three-tier course in the subject. The first year was devoted to "the dictionary of the science," the second to an initial unfolding of ideas behind it, and the third to acquisition of "true knowledge." As a student, Lavoisier found the program poorly organized. Later he wrote, "I was surprised to see how much obscurity surrounded the approaches to the science. During the first steps, they began by supposing in place of proving. They presented me with words which they were in absolutely no condition to define for me, or which at least they could not define without borrowing from knowledge that was absolutely foreign to me, and which I could only acquire by the study of the whole of chemistry. And so, in beginning to teach me the science, they supposed that I already knew it."

Here was just the sort of confusion that the mature Lavoisier loved to sort out and reorganize. And at the same time that he was enduring this lengthy but bewildering instruction in chemistry, he was also working in a much more clearly methodical fashion with the astronomer and mathematician Lacaille, in charge of mathematics and "exact sciences" (to which chemistry did not then belong) for the Collège Mazarin. Under Lacaille's tutelage, Lavoisier became, as he put it, "accustomed to the rigor of reasoning which mathematicians put into their work; never do they accept a proposition when the one which precedes it has not been discovered. Everything is connected, everything is linked, from the definition of the point, of the line, all the way to the most sublime truths of transcendant geometry." And Lacaille put as much emphasis on clear expression and precise terminology as he did on rigorous reasoning, exchanging the Latin that had been the traditional language of savants in any field for the French of his contemporaries—the Encyclopédistes, the Lumières, the scholars of the Enlightenment who were bringing their tongue to a point of refinement where it would replace Latin as the international language for any and all subjects.

After finishing his studies at the Collège Mazarin, Lavoisier took a series of chemistry courses at the Jardin du Roi from the famous Guillaume-François Rouelle—courses attended by many other emerging scientists of the French Enlightenment. Rouelle's chemistry was up-to-date, which in the 1760s meant that it was largely based on the theories of the German protochemist Georg Ernst Stahl. Rouelle had a relatively pragmatic, hands-on attitude to chemistry, which he defined as "an art which teaches us to separate by means of instruments, several bodies; to combine them to the end to be reacquainted with their properties and render them useful to the several arts." Techniques of analysis and synthesis were included in Rouelle's experimental practice, which as a gifted performer he knew how to dramatize for his audience of students, sometimes with an inadvertent explosion. Though progressive in the sense that he moved the study of chemistry in the direction of quantification, Rouelle also retained some aspects of alchemical philosophy, which he sought to combine with an emerging theory of empirically defined elements. The impossibility of this combination undoubtedly produced some confusion in his teaching.

"The celebrated professor," Lavoisier wrote of Rouelle, "joined much method in his manner of presenting his ideas to much obscurity in his manner of articulating them." This statement, surely as contradictory and confusing as whatever it meant to complain about, suggests that the student was simultaneously impressed and dissatisfied with the teacher and the course. Lavoisier's objection grew clearer as he went on: "I managed to gain a clear and precise idea of the state that chemistry had arrived at by that time. Yet it was nonetheless true that I had spent four years studying a science that was founded on only a few facts, that this science was composed of absolutely incoherent ideas and unproven suppositions, that it had no method of instruction, and that it was untouched by the logic of science. It was at this point that I realized I would have to begin the study of chemistry all over again." Here indeed was a radical statement; Lavoisier was willing to tear the rickety structure of chemistry right down to the ground and begin to rebuild it from point zero.

THE OBSCURITY OF chemistry as Lavoisier found it in the mid-eighteenth century owed much to its passage through medieval and Renaissance alchemy. Taken as a lump, alchemy is so far from the modern standard of exact science that twentieth-century writers like Joseph Campbell, Northrop Frye, and most notably the psychoanalytic philosopher Carl Jung have tended to treat it not as an exact science at all, but as a philosophical/religious system clothed in pseudochemical allegory; Jung indeed admired the architecture of the alchemical edifice, but only as an elaborate metaphor for processes in the human psyche.

Alchemy was certainly suffused with magical thinking, and its loftiest goals, such as the transmutation of base metals into silver or gold, the discovery or synthesis of the "philosopher's stone," and the acquisition of human immortality, were fantastically impracticable, at least on the plane of physical reality. Though modern science has shown the transmutation of metals to be possible after all, it turns out not to be worth the trouble and expense. A century before Lavoisier's time, one disillusioned alchemist denounced the project of transmuting metals as "a lunatic, melancholy fantasy."

Believing that it owned extremely valuable secrets, alchemy protected them in modes of expression that were most intentionally obscure. A parable from Bernard Trevisan's *De secretissimo philosophorum opere chemico* "begins with Bernard relaxing after a disputation by taking a walk in the open fields. He comes upon a beautifully constructed fountain and meets an old man there who tells him that the fountain's only use is as a bath for the king and that it is tended by a porter who warms it for him. Bernard asks the old man many questions about the king and his odd bathing practices and about the nature of the fountain. Eventually Bernard grows sleepy and accidentally drops a golden book (the prize from his disputation) into the fountain, drains the fountain to retrieve the book, and is thrown into prison for draining the king's fountain. After his release, Bernard returns to the fountain to find it covered in clouds."

George Starkey, a seventeenth-century alchemist, is sufficiently initiated in such arcana to be able to interpret this dreamlike allegory as a chemical formula. As described by historians William R. Newman and Lawrence M. Principe,

> First he [Starkey] notes that when the king, whom he easily identifies as gold, comes to bathe, he leaves "behind him all his servants (which are the metalls)" being accompanied only by a porter. Thus it seems here that iron, one of the lesser metals and thus one of the king's servants, must be "left behind." . . . This porter is surely the "Luna," or necessary medium between gold and mercury (i.e., the king and his bath), which he has identified as antimony regulus. Now Starkey notes that Bernard affirms "this Porter . . . to be most simple of al things in the world whose office is nothing but day by day to warme the bath (that is by making al fluid) now if it were compounded it could not be sayd to be so simple, which is uncompounded." Here Starkey interprets Bernard's use of the word "simple," which in the context of the parable means that the porter is unsophisticated or naive—"*homo valde simplex, imo simplicissimus hominum*"—to mean *compositionally* simple, that is, uncompounded, implying that pure "simple" antimony regulus should be jointed to the king/gold, not the regulus containing iron. . . . He next notes that Bernard asked the old man whether any of the king's servants went into

the bath with him and the "answer is returned not one, and if not one, then not iron."

Lunatic fantasy, indeed. By Lavoisier's day, several centuries' worth of this sort of baroque obfuscation had intervened between the scientists of the Enlightenment and the propositions about elemental chemistry made by the ancient Greeks. Much of the alchemist's time and effort went into the encoding and decoding of information that would prove scientifically dubious if ever it could be clearly understood. Carl Jung was not wrong in his contention that alchemy makes larger, more general sense if interpreted as a psychological/philosophical vision of humanity in the world than as an exact science, precisely descriptive and explanatory of material phenomena. Of course, this latter definition of *science* was just beginning to crystallize in Antoine Lavoisier's time. During the lengthy age of alchemy, and even during classical antiquity, the word *science* meant knowledge as a whole—knowledge of religion and philosophy as well as knowledge of concrete facts about the material world. When physics and metaphysics are not distinguished or divorced from one another, then mystical insight may be as legitimate a means of acquiring knowledge as any other—legitimate as the sort of experiment and analysis that fits the narrower definition of *science* we use today.

Jung, in an essay on Paracelsus, noted that "the alchemist . . . worked alone. . . . This rigorous solitude, together with his preoccupation with the endless obscurities of the work, was sufficient to activate the unconscious and, through the power of imagination, to bring into being things that apparently

were not before." Though Jung does not mean it so pejora-
tively, this statement bears a certain resemblance to
Lavoisier's analysis of the role of imagination in mesmerism.
Paracelsus himself put the point more plainly and more pos-
itively: "Magic has the power to experience and fathom
things which are inherently inaccessible to human reason.
For magic is a great secret wisdom, just as reason is a great
public folly."

A magically intuitive fathoming of order and organization
in the universe is the basis of alchemy's appeal as a mode of
religious philosophy. It was unfortunate that alchemy's most
elegant philosophical and psychological insights kept falling
short of the standards of material proof—but for centuries
alchemists had a reflexive defense: "In alchemy, the laboratory has no crucial role. The function of the practice is first
and foremost to illustrate the truth of the theory. The success
of a procedure demonstrates to the operator that he has
understood the ancients well. The quality of the practice is
the direct consequence of the level of understanding of the
theory. For if the experiment fails, the failure does not weaken
the theory."

In other words, alchemical theory was analogous to religious faith in its defiance of logic (when necessary) and in its
imperviousness to inconvenient facts. And in this respect, the
relationship of alchemical theory to alchemical practice was
almost a perfect inversion of modern scientific method. For
modern chemistry to be born, the alchemical world had to be
turned upside down. In the seventeenth century, that world
was already beginning to tilt.

Seventeenth-century alchemy was not all mystical hocus-
pocus—permeated with magical thinking as it continued to

be. Transitional figures such as Starkey and Robert Boyle continued to draw on alchemical lore, but subjected it to genuine experimental testing. Boyle, much influenced by the deductive methods of Francis Bacon, set forth in his treatise *The Sceptical Chymist* a much more empirical approach to chemistry than had previously been seen. Instead of disregarding experimental results when they were incompatible with a preexisting theory, Boyle and his ilk saw "failed" experiments as a challenge to existing theory—as doorways to new experimental programs and, accordingly, revision of theory. At the same time, he persisted in believing that transmutation of metals was practically possible and that the philosopher's stone would one day be discovered. Because of the fundamental differences between the practice of Boyle, Starkey, and Jan Baptista van Helmont and that of both earlier alchemists and modern chemists, historians Newman and Principe distinguish the transitional seventeenth-century science as "chymistry."

ALONGSIDE ALCHEMY EXISTED a parallel tradition related to the mining and refining of metals, codified in a manual of metallurgy by Georgius Agricola in the mid-sixteenth century. The metallurgical tradition was opposite to alchemy in its attitudes and intentions and so rather closer to those of modern science. Miners and refiners had a practical interest in creating a clear and accessible body of knowledge about their craft. Their practices were firmly based on results. Procedures needed to work reliably and be repeatable by anyone trained in them. Terminology had to be consistent and clear. The encryption of alchemical texts was irritating to Agricola, who wrote that "all are difficult to follow, because the writers upon these things use strange names, which do not properly belong to the metals, and because some of them employ now one name and now another, invented by themselves, though the thing itself change not."

The metallurgical tradition produced a much more reliable lexicon of metals and minerals than alchemy ever wanted to do; Lavoisier himself relied on it when conducting his mineralogical expeditions with Guettard. And where alchemy was secretive and intentionally arcane, the techniques of mining and metallurgy were relatively open, since they had broad economic significance in the communities where they took place. But despite their very different attitudes, alchemy and metallurgy had a common interest in precious metals and used similar furnaces, crucibles, and distillation apparatus. Both traditions were engaged in the refinement and purification of elements by fire.

THE MODERN PERIODIC table identifies more than one hundred elements; its prototype, the Table of Chemical Nomenclature published by Lavoisier and his colleagues in 1787, lists fifty-five. Lavoisier's table constituted a radical change in the whole notion of what an element was. Prior to his reorganization of the definitions and concepts, the Western scientific community had continued to labor over revisions of the elemental theory inherited from the ancient world.

The ancient elements were defined according to their direct accessibility to the senses—without any operations of analysis or chemical decomposition. The Chinese system identified five elements—metal, wood, fire, water, and earth—derived from the tension of opposites described in Taoism. Metal and fire were seen as yang elements—hot, bright, and masculine—while wood and water were yin elements—cool, dark, and feminine. The earth element stood on neutral ground, centered between the extremes of yin and yang. Compound substances were understood according to the proportions of the five elements they contained.

Chinese five-element theory was the basis for Chinese alchemy, which cross-pollinated with Western alchemy to some degree and which, like Western alchemy, had an interest in the transmutation of metals and the discovery of the secrets of immortality. Five-element theory also was and continues to be the basis for Chinese medicine, which evolved on a very different course from its Western counterpart. Chinese medicine still produces therapeutic results for patients all over the world, while for the most part remaining untouched by the logic of Western science.

Circa 450 B.C., the Greek philosopher Empedocles proposed four elements: fire, earth, air, and water. Aristotle supplemented the elements with four "qualities": defining fire as hot and dry, water as cold and wet, air as hot and wet, and earth as cold and dry. As in Chinese five-element theory, compound substances were understood as mixtures of the four Aristotelian elements and the degrees of their attendant qualities. Nature conferred qualities to combinations of the elements to produce the metals mined from the earth. Western alchemy believed that this natural process could be artificially replicated, and so conceived the philosopher's stone as the mechanism for the imposition of metallic qualities on "prime matter."

During the Renaissance, Western alchemy intertwined itself with the tradition of Hermeticism and Hermetic beliefs in quasi-magical correspondences between the macrocosmos and the microcosmos, so that, for example, the macrocosmic organization of the universe could be mapped onto the microcosmic organization of the human body. Similarly, the metals were mapped to corresponding bodies of the ancient planetary system: lead to Saturn, copper to Venus, iron to Mars, silver to the moon, gold to the sun, and so on. Via this connection, the chemical vocabulary began to derive its terms from astrology, and the astrological idea of planetary influence took on an importance for alchemy.

Both Western and Chinese alchemy were interested in the creation of health along with wealth; alchemy contained a thread of nascent pharmacology up through seventeenth-century chymists such as Starkey and Boyle, who earned something from drugs and their formulae. Western medicine derived its fundamentals from the Greek Galen, who proposed four humors analogous to the Aristotelian four elements and defined states of health or illness according to the balance of humors in the body. Paracelsus, who directed his alchemy more toward medicine than metallurgy, rebelled against the Greek system of humors and elements and declared the existence of three primary principles—sulfur, mercury, and salt—which corresponded both to the Holy Trinity and to the components of the human being: "vital spirit, soul, and body." According to Paracelsus, all alchemical transformation was controlled by the interaction of this *tria prima*.

As a doctor, Paracelsus had a useful strain of empiricism (he was, for example, the first to identify the causes of silicosis, a miners' lung disease), but on the theoretical plane he remained a metaphysician and (by his own declaration) a magician. His rebellion did not succeed in dislodging Aristotelian science from orthodox European thought, but his ideas did have a pervasive influence throughout the seventeenth century. In what amounts to a prototypical vision of organic chemistry, Paracelsus saw life processes as a mode of alchemy; God, as Creator, is "the supreme alchemist." Paracelsus and his followers objected to Aristotelian definitions (in sixteenth-century botanical nomenclature, for example) *because of* their descriptive and empirical character, which, in the Paracelsian view, failed to capture the universal correspondences on which authentic definitions must depend. "Within this vast and vital universe, the *true* physician had to reveal the hidden relations between microcosmos and macrocosmos, and interpret the signatures concealed by God in each single body."

Before dismissing Paracelsus as a lunatic fantasist, it is instructive to compare him to a true physicist, Isaac Newton, who studied alchemy throughout his career, during the same decades that he explained gravitation and other fundamentals of physics. The laws of Newtonian physics (in the eyes of Newton, at least) were originally God-given. Like Paracelsus, like most alchemists in fact, Newton saw himself as a discoverer of divine properties installed by God in the natural world. His physics was thus subordinated to metaphysics, and his view of the universe as holistic as that of the philosophers, alchemists, and mystics. Lavoisier, though impressed by Newton and influenced by the logical rigor of Newtonian physics, would begin to deconstruct this holistic vision of the universe by concentrating much more narrowly on its component parts.

ROBERT BOYLE, A COUNTRYMAN and colleague of Isaac Newton's, was not so successful in rationalizing chemistry as Newton was in rationalizing physics, though he did have an interest in doing so. In *The Sceptical Chymist*, he challenged both the Aristotelian and the Paracelsian conceptions of the elements. Seventeenth-century chymists had begun to suspect that the Aristotelian elements were mixtures, rather than pure substances. Van Helmont, another chymist whose work Boyle knew, contended that water contained both mercury and sulfur in its composition—two elements of Paracelsus's *tria prima*.

But Boyle also undermined the *tria prima* via analysis by fire. Through various experiments he concluded "That the Fire even when it divides a Body into Substances of divers Consistences, does not most commonly analyze it into Hypostatical Principles, but only disposes its parts into new Textures, and thereby produces Concretes of a new indeed, but yet of a compound Nature." Boyle invites his reader to consider, as a mode of analysis, "the Burning of Wood, which the Fire Dissipates into Smoake and Ashes: For not only the latter of these is Confessedly made up of two such Differing Bodies as Earth and Salt; but the former being condens'd into that Soot which adheres to our Chimneys, Discovers itself to Contain both Salt and Oyl, and Spirit and Earth, (and some portion of Phlegme too) which being, all almost, Equally Volatile to that Degree of Fire which Forces them up, (the more Volatile Parts Helping, perhaps, to carry up the more Fixt ones, as I have often Try'd in Dulcify'd *Colcothar*, Sublimed by *Sal Armoniack* Blended with it) are carried Up together, but may afterward be Separated by other Degrees of Fire, whose orderly Gradation allowes the Disparity of their Volatileness to Discover itself."

In fewer words, the decomposition of wood by fire yields more compound substances, rather than pure, elementary ones. Meanwhile, Boyle recorded that pure gold would not decompose no matter how much it was heated, and that if he heated "a Mixture of Colliquated Silver and Gold . . . in the Fire alone, though vehement, the Metals remain unsevered," though they could easily be separated "by *Aqua Fortis*, or *Aqua Regis* (according to the Predominancy of the Silver or the Gold)." Boyle's distillation of blood yielded "phlegm, spirit, oil, salt, and earth"—five substances that were not certainly elementary. The chymical skeptic also distilled eels by boiling them and determined that "they seem'd to have been nothing but coagulated Phlegm, which does likewise strangely abound in Vipers."

The very inconsistency of Boyle's results—the product of his empirical skepticism—invalidated the existing elemental theories, since analysis by fire failed to decompose different substances into the same set of primary elements. Moreover, Boyle had begun to discern that "the Fire may sometimes as well alter Bodies as divide them, and by it we may obtain from a Mixt Body what was not Pre-existent in it"—that is, combustion could form new compounds with components not present in the substance being burned.

THE GERMAN PHYSICIAN and chemist Georg Stahl, whose career spanned the turn of the eighteenth century, built on a transformation of Paracelsian elemental theory by an earlier German experimenter, Johann Joachim Becher. In place of Paracelsus's *tria prima*, Becher adopted air, earth, and water as the three elementary principles; for the moment fire, the fourth Aristotelian element, was removed from the picture. Becher further supposed that three different kinds of earth were required for the composition of metals and minerals, and that one of these earths, *terra pinguis* or "greasy dirt," contained the principle of combustion. Van Helmont (a follower of Paracelsus and in some respects a leader of Becher) had earlier used the Greek *phlogistos* for inflammability; Becher employed the same term; Stahl adapted it to "phlogiston."

Newton's admirers hoped and expected that chemical questions would be mechanically resolved "in terms of the interaction of matter and *forces*," as Newtonian physics resolved physical questions. Stahl disagreed, distinguishing (as Becher had done, but with greater refinement) between mixtures that he called "aggregates" and "mixts." An *aggregate* was a mechanical mixture of substances accomplished by physical forces—like grains of sand shaken up in a jar. A *mixt*, by contrast, required a chemical reaction to produce it and therefore was a genuine chemical compound. Stahl defined *chemistry* as "the art of dissolving natural mixt bodies by various means"—that is, the analysis of compounds.

Fire, in Stahl's chemistry, was not an element but an instrument; fire was not an ingredient to any mixt but rather the mechanical tool whose action assisted the mixt into being. Becher's *terra pinguis*—Stahl's phlogiston—was the material ingredient acted on by fire in Stahl's theory of corrosion, combustion, and calcification.

Stahl thought of rusting as a slow-motion version of burning. He theorized that when a metal rusted, or when any combustible substance burned, it lost a portion of the phlogiston it was supposed to contain. An analogue to these processes was calcination, where heating of some metals produced what modern science calls an *oxide* and emergent eighteenth-century chemistry called a *calx*. Such calces were chemically identical to raw ores mined from the earth. According to Stahl's theory, the refinement of metals by smelting ores with charcoal (a practical procedure long and well known in the mining and metallurgy tradition) involved the transfer of phlogiston from the charcoal to the ore, which took on phlogiston to become a refined metal. In calcination, when heating degraded metals into ores, the metals supposedly lost phlogiston to the surrounding atmosphere.

The theoretical phlogiston was the "matter of fire," the particular "sulfurous earth" that accounted for combustibility. Materials that burned readily, such as wood, charcoal, or sulfur itself, did so because they were rich in phlogiston. Modern chemistry understands that flames die in sealed spaces when all the available oxygen has been consumed. Stahl explained this phenomenon in reverse: flames died when the burning material had released all its phlogiston, saturating the surrounding air to the point that it would no longer support combustion. Further, Stahl reasoned that phlogiston released into Earth's air by burning was reabsorbed by plants and trees—thus, wood acquired the phlogiston, which made it so highly combustible.

The phlogiston theory was wrong, but it worked. It had the fundamental scientific virtue of accounting for a wide range of empirical observations with a single, self-consistent explanation. For that reason, most chemists of the mid-eighteenth century had grown firmly attached to it, and indeed it had been installed as part of the common knowledge of educated people. In his *Critique of Pure Reason*, Immanuel Kant appreciated Stahl's theory as a milestone in the progress of science: "When Galileo caused balls, the weights of which he had himself previously determined, to roll down an inclined plane; when Torricelli made the air carry a weight which he had calculated beforehand to be equal to that of a definite volume of water; when Stahl changed metals into oxides, and oxides back into metals, by withdrawing something and then restoring it, a light broke upon all students of nature."

Stahl's chemistry was state of the art in the early 1760s, when Lavoisier was a student. But Lavoisier got his first definition of phlogiston from Rouelle, who had somewhat modified Stahl's idea. In Stahl's system, phlogiston was a "principle" that entered into the composition of mixts, while fire itself was an instrument that operated externally on the formation of mixts without entering into them. Rouelle's chemistry course identified fire and phlogiston more fully with one another: "We recognize four elements: phlogiston or fire, earth, water and air." At the same time, some of Rouelle's precepts were 100 percent alchemical; for example, "The philosopher's stone is nothing else but the result of the fermentation of gold with mercury especially charged with phlogiston. That is all I have to say of those transmutations which so many people without knowledge talk about."

Lavoisier found Rouelle's instruction both admirable and frustrating; he had recognized a need to "begin the study of chemistry all over again" when in 1766 he purchased one of Stahl's Latin manuscripts—a treatise on sulfur. Lavoisier's numerous annotations concentrated on the sections to do with calcination and combustion, on which Stahl's phlogiston theory was based. In his direct study of Stahl, Lavoisier found for the first time a systematic, empirically grounded theory of chemistry. Though his own work would finally invalidate Stahl's notion of phlogiston, Lavoisier understood the usefulness of Stahl's concept: "For the first time in the history of chemistry, a theory was embodied in the facts it aimed to explain."

———————

IN 1793, AT around the same time that Lavoisier, haunted by the ghost of Marat, came under suspicion for his activities with the General Farm, Antoine-Nicolas de Caritat, marquis de Condorcet, went underground in the Paris house of a friend, Madame Vernet. Though lately a colleague of Robespierre and a member of the Committee of Public Safety, Condorcet had recently made himself a target of the Terror by publicly protesting the Jacobin Constitution adopted earlier the same year. He passed the nine months he spent in hiding *chez* Madame Vernet by drafting *The Sketch for a Historical Picture of the Progress of the Human Mind*, a brief but sweeping survey of human development from prehistoric times to his own rapidly contracting present. The work complete, Condorcet emerged from his refuge, was promptly arrested, and soon died in prison, perhaps by suicide.

From a perspective not far from Lavoisier's, Condorcet describes the broad influence of Newton: "We owe to Newton and to Leibniz the invention of these calculi for which the work of the geometers of the previous generation had prepared the way. . . . When we come to describe the formation and the principles of the language of algebra, the only really exact and analytical language yet in existence, the nature of the technical methods of this science and how they compare to the natural workings of the human understanding, we shall show that even though this method is by itself only an instrument pertaining to the science of quantities, it contains within it the principles of a universal instrument."

Following an account of Newton's mathematical explication of the law of gravity, Condorcet adds that "Newton perhaps did more than discover this general law of nature; he taught men to admit in physics only precise and mathematical theories, which account not merely for the existence of a certain phenomenon but also for its quantity and extension." In these passages Condorcet is looking back—just over his shoulder—at the Enlightenment project of applying Newtonian methodology not only to transform all branches of the old natural philosophy into exact, mathematicized sciences, but also to reform all other branches of human knowledge—politics, metaphysics, history itself—on a similar basis. This impulse accounted for the creation of the metric system (on which Lavoisier was laboring while Condorcet wrote), as well as the more dubious achievement of the French Revolutionary Calendar, which, at the time of Condorcet's writing, had tried to take history back to zero.

Before Lavoisier completed his fundamental work, chemistry was a Baconian rather than a Newtonian science—a vast anthology of facts (and pseudofacts) not very well marshaled by theory. Even in Lavoisier's grip, chemistry would prove resistant to Newtonianization and to complete mathematicization as well. But Lavoisier was early to acquire the idea of modeling a new approach to chemistry on the huge advances that had been made in experimental physics; indeed, he did not completely distinguish the two disciplines, and in the beginning considered himself a physicist as much as a chemist. In 1766, during the early phase of his siege of the Academy of Sciences, he submitted an argument for increased emphasis on physics in the Academy: "experimental physics has escaped from the shadowy laboratories of earlier chemists* and . . . begun to take on a new form. Firmly based on experiments and facts, it has steadily advanced." Already,

* One is reminded of Boyle, with his cauldrons of vipers.

Lavoisier intended to drive the shadows farther from the laboratories by establishing chemistry on a similarly firm base.

Kant, who like Condorcet regarded mathematics as the "universal instrument," thought that all "genuine science" must have a mathematical foundation, and argued in his *Metaphysical Foundations of Natural Science* that chemistry could not aspire to the condition of genuine science *because* it relied on collections of empirical, Baconian facts, rather than proceeding from theoretical axioms. By contrast, Lavoisier followed his teacher Lacaille in believing that mathematics itself derived from the quantification of observations about the natural world, and so should be understood as "a highly formalized way of stating empirically based knowledge."

In the early 1760s (soon after the death of his mentor Lacaille), Lavoisier attended lectures given by Jean-Antoine Nollet, a physicist whose Cartesian approach to the subject would soon be eroded by his colleague and rival in the Academy of Sciences, George-Louis de Buffon, who adopted the Newtonian stance. Lavoisier got his grounding in the methods of experimental physics from Lacaille and Nollet, and during the years that he himself was striving for admission to the Academy of Sciences, the Nollet-Buffon controversy gave him his first full demonstration of the politics involved in the acceptance or rejection of new scientific ideas.

In the end, Nollet got the worst of it. Buffon, having recognized that Benjamin Franklin's discoveries about electricity superseded Nollet's ideas on the subject, used a French translation of Franklin's work to undermine Nollet's reputation, and soon was able to declare Nollet to be "dying of chagrin from it all." Contentiousness over the issue actually brought electrical research in France to a deadlocked halt, a point that the progressive Lavoisier would not have failed to notice. Meanwhile, Buffon pushed past Nollet in his promotion of a Newtonian chemistry, which tried to account for chemical affinities as a sort of miniaturized version of the law of gravitation.

Lavoisier was, in general, suspicious of this sort of retreat into theoretical abstraction. His first sketch of a reformed course in chemistry, drafted when he was twenty-one, was modeled on Nollet's course in physics—replete with experimental demonstration. Both Nollet and Lacaille were skilled designers of scientific equipment, and Lavoisier's strong interest in precision instruments and the fine quantification they made possible began in his experience with these two teachers—here was one of the first features of experimental physics that he determined to export to chemistry. His notion of the application of mathematics to the quantification of scientific data also originated with Lacaille.

Scientific theory had to emerge from interpretation of precisely quantified data—as mathematics itself, in Lacaille's view, derived from the organization of empirical information. "The only way to prevent errors," Lavoisier wrote in his preface to the *Traité élémentaire de chimie*, "is to suppress reason, or at least simplify it to the greatest extent possible, for it comes entirely from us and if relied on can mislead us." The suppression of reason may seem an odd enterprise for a luminary of the eighteenth-century Age of Reason, but what Lavoisier meant was to stress the importance of cross-examining theory whenever it drifted away from demonstrable fact. Such, for example, was the method he used to debunk mesmerism: driving a wedge between the mesmerists' theory and the facts they claimed that it explained.

"Reason must continually be subjected to experimental

proof. We must preserve only those facts that are given by nature, which cannot deceive us. Truth must only be sought in the natural connection between experiments and observations, in the same way that mathematicians arrive at the solutions of a problem by a simple arrangement of the givens. By reducing reason to the simplest possible operations and restricting judgment as much as possible, they avoid losing sight of the evidence that guides them." Empirical facts, in Lavoisier's methodology, would always be primary. Experimental demonstration, modeled on the rigor of geometric proof, would build those facts into a durable structure of ideas.

LAVOISIER WAS AS meticulous as Sherlock Holmes in the close examination he gave to his experimental proof. The exactitude of measurements assumed a paramount importance. For that reason, Lavoisier took an acute interest in refining the subtlety and the accuracy of his laboratory equipment.

Early in the seventeenth century, the Belgian chymist van Helmont had performed an experiment that many eighteenth-century scientists still believed to be a demonstration of the transmutation of water into earth. Van Helmont potted a 5-pound willow in 200 pounds of soil, closed the soil container, and added nothing but rainwater. At the end of five years the weight of the soil was unchanged, but the willow had increased to a weight of 169 pounds. The role of photosynthesis in plant growth was nowhere near discovery in van Helmont's day, nor had the phlogiston theory (which Stahl did use to account for plant growth) yet made its appearance. Van Helmont reasoned that the water must have been transmuted into earth in order to generate the increased wood of the tree.

Though the transmutation of metals was already discredited by the time of Lavoisier's early experiments, the notion of transmutation of other substances persisted as a corollary of the Aristotelian theory of the elements, which had not been definitively replaced. Proceeding from the theory that water could be transmuted into earth, scientists following van Helmont explained the observation that distilled water always left a solid residue in the vessel in these terms.

In conjunction with his studies of the public water supply in the late 1760s, Lavoisier became interested in the transmutation question. He suspected that the solid residue left by distillation more likely came from glass dissolving from the vessel during boiling. To prove the point, he boiled three pounds of water for one hundred days in a glass vessel called a "pelican" (its curved, hollow handles, which worked as distillation tubes, resembled the wings of the bird). At the end of the experiment Lavoisier found that there was indeed a residue, but its weight was almost exactly equal to the weight lost by the pelican. The sum of the weights of the pelican and its contents had not changed. The shift of weight from the vessel to the contents was accounted for by grains of salt in the residue. Lavoisier's hypothesis—that material dissolved from the vessel furnished the solid residue of distillation, rather than any transmutation of one element into another—was so demonstrated.

This early experiment had a minor flaw: while the weight lost from the pelican was 12.5 grains, the weight of the salts in the residue was 15.5 grains. Lavoisier permitted himself to overlook the discrepancy, or rather (somewhat contrary to his methodological rhetoric) he allowed his governing theory to overrule it.

The theoretical postulate involved was the conservation of

matter, also known as the *conservation of mass*. Lavoisier did not formally express it until 1785: "Nothing is created either in the operations of art, or in those of nature, and it may be considered as a general principle that in every operation there exists an equal quantity of matter before and after the operation; that the quality and quantity of the constituents are the same, and that what happens are only changes, modifications. It is on this principle that is founded all the art of performing chemical experiments; in all such must be assumed a true equality between constituents of the substances examined, and those resulting from their analysis."

Though Lavoisier generally gets credit for the authorship of this principle, others had conceived it before him. The seventeenth-century chymists, notably van Helmont, Starkey, and Boyle, had a dawning awareness of the importance of weighing and measuring materials before and after an experimental process, though their methods and their measurement devices were not so precise. In 1623, Francis Bacon declared, "Men should frequently call upon nature to render her account; that is, when they perceive that a body which was before manifest to the senses has escaped and disappeared, they should not admit or liquidate the account before it has been shown to them where the body has gone to, and into what it has been received." And as early as 450 B.C., Anaxagoras argued, "Wrongly do the Greeks suppose that aught begins or ceases to be; for nothing comes into being or is destroyed; but all is an aggregation or secretion of preexisting things; so that all becoming might more correctly be called becoming mixed, and all corruption, becoming separate."

Anaxagoras's "nothing comes into being or is destroyed" is very close indeed to Lavoisier's "nothing is created." The idea

of conservation of matter had been around for many centuries before Lavoisier installed it as the centerpiece of his
experimental method. (He depended heavily on the principle
for fifteen years before he announced it, perhaps because he
simply assumed that its validity was common knowledge.)
But Lavoisier, beginning with this experiment in the distillation of water, deployed the principle much more strictly and
consistently than any scientist had ever done before him.

Lavoisier's work in finance continually reinforced his com-
mitment to the idea of balance. In all phases of his career, he
was an exacting accountant. An analogy to physics encouraged
him to view money as "a fluid whose movements necessarily
end up in a state of equilibrium." Equilibrium governed the
balance scale on which the materials of chemical experiments
were weighed, and increasing the refinement and the accuracy
of this instrument was for Lavoisier a perpetual concern.

BY THE EARLY 1770S, Lavoisier's preliminary investiga-
tions had progressed halfway through the sequence of four
Aristotelian elements. The mineralogical surveys he conducted with Guettard broadly covered the earth element. His
studies on behalf of the Academy of Sciences of the Parisian
water supply and the general properties of water covered the
element of water as thoroughly as could be done at the time.
Now Lavoisier turned to the study of air. More work in pneumatic chemistry—the chemistry of gases—had been done in
England than in France at that point. Several different gases
present in the atmospheric air had been isolated, though with
no terminology to identify them and no precise understanding of what they were. The various cloudily identified gases
were still considered to be inert rather than reactive in chem-

ical combinations. Georg Stahl, whose chemistry remained the most advanced on record, believed that air was merely an environment surrounding chemical reactions, not an active ingredient in them.

In France the 1770s saw a resurgence of a practice that had appeared a century before in the court of Cosimo III, grand duke of Tuscany: a scientific fad for incinerating diamonds. The Tuscan experiment subjected six thousand florins' worth of diamonds and rubies to twenty-four hours of extreme heat and found that while the rubies were unaffected, the diamonds had vanished without a trace. In the 1770s, French chemists, Lavoisier among them, took up this ostentatiously extravagant line of research.

The equipment involved a huge contraption, vaguely resembling a Roman catapult, that focused sunlight via two enormous lenses, called "burning glasses," to concentrate intense heat on the crucibles of jewels. The scientific operators wore smoked goggles to protect their vision from the brilliance of the burning. This engine was wheeled into the Jardin de l'Infante, outside the Academy of Sciences quarters in the Louvre, and close beside a popular public promenade along the Seine. Onlookers were plentiful, and the ladies were impressed (and perhaps dismayed). Of course, there was much more dramatic interest in the burning of diamonds than if the scientists had been applying their incendiary beam to ordinary lumps of coal.

It was soon determined that the presence of air was required for the diamonds to disappear or be consumed. Diamonds effectively sealed from the atmosphere were always recovered intact. Today we know that since diamonds are a form of carbon, sufficient heat makes their carbon combine with oxygen and vanish into carbon dioxide. Lavoisier must have suspected that the diamonds were disappearing into gases, for in the spring of 1773 he conducted a series of experiments in which he tried to capture any gases that might be emitted from burning diamonds under a glass bell. But his glass vessels shattered in the extreme heat of these experiments, the gases could not be collected for measurement, and the cause for the dissolution of diamonds could not be firmly defined.

THEN LAVOISIER TOOK note of a more promising line of investigation in pneumatic chemistry—one involving the calcination of metals. Stahl's phlogiston theory asserted that metals, when heated to form calces, released or lost phlogiston. Why, then, did the resulting calces weigh more than the original quantities of metal that went into their formation? Phlogiston was supposed to have a weight (though as it did not really exist, no one had so far managed to weigh it). So the increased weight of a substance supposed to have lost phlogiston violated the principle of conservation of matter—which to Lavoisier was absolutely axiomatic.

Because of this inconsistency, explanations of calcination in terms of phlogiston grew more and more tortured, yet Condorcet was articulating the prevailing view when he wrote, in reaction to the challenge Lavoisier was beginning to formulate, "If ever there was anything established in chemistry, it is surely the theory of phlogiston." Turgot, alongside his career as economist and Physiocrat, was also accomplished enough as an amateur chemist to be invited by Diderot to contribute chemical articles to the French *Encyclopédie*. "The increase in weight occurring in metal," he wrote in his entry on the weight gain of metallic calces, "is due to the air which,

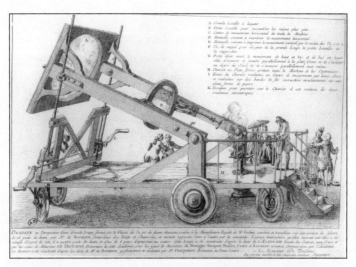

The "burning glasses," capable of producing intense heat.

in the combustion process, combines with the metallic earth and replaces the phlogiston, which is burned and which, without being of an absolute lightness, is incomparably less heavy than air, apparently because it contains less matter." This passage is a difficult one. Phlogiston, now that it needed both to have weight and to be lighter than air, was well on the way to becoming a *deus ex machina* of chemical reactions—not to say a magical solution.

The exacting mind of Lavoisier was quick to home in on small weaknesses of this sort. In February of 1773, he opened a new laboratory notebook with a plan for a research program in this area:

Before beginning the long series of experiments that I propose to myself to do on the elastic fluid which releases from bodies, be it by fermentation, by distillation, or finally by all kinds of combinations, as well as on the air absorbed in the combustion of a great number of substances, I believe I ought to put down some reflections here in writing, to form for myself the plan that I must follow.

However numerous are the experiments of MM. Hales, Black, Mac Bride, Jacquin, Crantz, Priestley, and de Smeth on this subject, it nevertheless remains necessary that they should be numerous enough to form a complete body of theory.... The importance of the subject has engaged me to again take up all this work, which strikes me as made to occasion a revolution in physics and chemistry. I believe that I should not regard all that has been done before as anything other than indications; I propose to repeat everything with new precautions, so as to connect what we know about air which fixes or releases itself from bodies with other acquired knowledge and to form a theory.

The works of the different authors which I have just cited, considered from this point of view, have presented to me separate sections of a great chain; they have joined together a few links of it. But a great sequence of experiments remains to be done, to form a continuity.

Despite his declared suspicion of theory, Lavoisier's most durable achievements would be as a theoretician—not as a discoverer of previously unknown facts. In 1773, when he made these notes, he already seemed to know it.

III

Le principe oxygine

Though Lavoisier is generally credited with the discovery of oxygen, he was not the first person to isolate the gas. A surprising number of other scientists had done so—a surprisingly long time before he did. Lavoisier's real and significant "first" was to identify oxygen as such—and to define oxygen in terms of a theory embodied in its facts.

The Academic structure of the European intellectual community of the eighteenth century fostered an increased awareness of individual responsibility and credit for new discoveries. The discoveries of the alchemical world, such as they might have been, were never so loudly or lucidly announced; the metaphorical encryption of alchemical texts was a way of protecting the knowledge. Seventeenth-century chymists were open in the presentation of their results and interpretations; they also plagiarized each others' work with an abandoned enthusiasm.

Lavoisier was as sensitive to the increasing importance of priority in scientific discoveries as any of his colleagues, in

France or abroad—and perhaps more so than most. Thus, even before predicting the revolutionary significance of his research program in his laboratory register in February of 1773, Lavoisier took steps to protect the priority of his discoveries, somewhat in advance of actually making them.

The French phrase *pli cacheté* might be translated (fancifully) as "hidden wrinkle"; more accurately, it was the sealed note that Lavoisier deposited with the secretary of the Academy of Sciences on November 1, 1772. Experiments he had undertaken in October of that year confirmed that burning phosphorus increased in weight as it formed phosphoric acid, and that the calcination of sulfur produced a similar weight gain. Such increases in weight had been observed before, but usually were accounted for by torturing Stahl's phlogiston theory. To Lavoisier it made little sense that a substance would gain weight by losing phlogiston (or anything else); it was more likely, he began to hypothesize, that sulfur, phosphorus, and other substances were absorbing something, rather than releasing it, during calcination and combustion, and that something must be air or some component of the air.

He labeled the sealed note "On the cause of the weight gained by metals and several other substances when they are calcined." In the interior he scrawled (evidently in some haste, given his numerous cross-outs and the vagaries of his spelling and syntax):

It has been about eight days since I have discovered that Sulphur while burning far from losing any of its weight ~~while burning~~ acquires some on the contrary; That is to say that from one pound of Sulphur one could extract much more than one pound of vitriolic acid, the extraction done in the humidity of the air.* it is the same with phosphorus. This augmentation of Weight comes from a prodigious quantity of air which fixes Itself during the combustion and which Combines Itself with the vapours.

This discovery which I have Confirmed by some experiments which I regard As decisive has made me think ~~what~~ that What is observed in the Combustion of Sulfur and of Phosphorus could very well take place with regard to all Bodies which acquire weight by Combustion and Calcination and I Am persuaded that the augmentation of weight ~~of the~~ of metallic chalks holds the same Cause. the experiment has Completely Confirmed my Conjectures. i have made the reduction of litharge in closed vessels with the apparatus of M. hales and I have observed that at the moment of passage from calx to metal there ~~is produced~~ disengages a Considerable quantity of air and which forms at least a volume a thousand times greater than the quantity of litharge used. *This discovery seems to me one of the interesting to have been made since Stahl and As it is difficult not to let slip to One's friends in Conversation something which could put them on the way to the truth I Believed I should make the present deposit into the hands of M. Secretair of the academy for while waiting to make my experiments public.*

Lavoisier was not quite thirty when he scribbled these notes, just moving from youth into his mature prime. He had won Academy membership on a provisional basis only four years before, and was eager to consolidate his position with

* Combustion of sulfur in oxygen forms sulfur dioxide: $S(s) + O_2(g) >> SO_2(g)$.

some important scientific revelation, as the last lines of the sealed note clearly show. Undoubtedly he was aware that other scientists, not only on the European continent but also in England, had become acutely interested in the role that air played in combustion and calcination.

Modern science has established that atmospheric air is composed of approximately 80 percent nitrogen, approximately 20 percent oxygen, and less than 1 percent each of various other gases, such as argon, methane, carbon dioxide, krypton, ammonia, and so on. Lavoisier's theory, and the key experiments that supported it, opened the door to this analysis. Prior to Lavoisier's work, the conception of atmospheric air was sometimes simpler, sometimes not, but never nearly so accurate.

In the Aristotelian system, air was one of the four elements and as such was understood to be irreducible. Paracelsus discarded the four-element theory in favor of his *tria prima* of mercury, sulfur, and salt. Van Helmont refined the Paracelsian proposition into what might be called a *one-element theory*, in which water was the fundamental principle and the base of all material transmutations. Becher, who extended Paracelsian ideas in a different direction than van Helmont, replaced the *tria prima* with air, earth, and fire, but considered air to be not an element but an agent or instrument for chemical mixtures. The seventeenth-century chymists elaborated these notions to the point that by the early eighteenth century the generally accepted doctrine was less an elemental theory than an instrumental one in which water, air, and fire were understood to be agents, rather than components, of physical change.

Georg Stahl, whose chemical theories were dominant at the time when Lavoisier came onto the scene, thought of atmospheric air as chemically inert, incapable of entering into chemical combinations. Hermann Boerhaave, though he differed from Stahl in many respects, shared his conception of air in the beginning, but later changed his mind. Or rather, he had it changed for him by an Englishman, the Reverend Stephen Hales.

Hales was both a botanist and a chemist; his *Vegetable Staticks*, the book so influential on changing the eighteenth-century chemists' attitude toward air, was mainly concerned with the application of Newtonian principles to plant life, as the title suggests, but *Vegetable Staticks* also included a chapter called "Analysis of Air." Hales's observation that plants absorbed and somehow processed large amounts of air prompted him to study air as a thing in itself, and his impressions were somewhat contrary to the received ideas of his time. Whereas Stahl had declared that atmospheric air entered into no chemical compounds ever, Hales concluded from experiments in which he measured air released in processes like fermentation and distillation that "fixed" air was a component of all organic materials and some inorganic ones.

In the official edition of his *Elementa chemiae* (a 1732 volume replacing a pirated collection of his lecture that had surfaced eight years previously), Boerhaave dropped Stahl's conception of air in favor of Hales's, which meant, effectively, departing from the theory of four instruments to return in the direction of the old four-element theory. Lavoisier's chemistry teacher, Rouelle, attempted his own synthesis of the ideas of Boerhaave, Hales, and Stahl. Rouelle transmitted many of Stahl's concepts and was responsible for popularizing them in France, but he departed from Stahl by teaching that earth, water, fire, and air were not simply instruments, in Stahl's sense, but also chemically active. Rouelle argued that air was a component fixed in matter, and demonstrated the point (as Hales had done, and often using equipment Hales had designed) with experiments that measured air released from various substances by distillation, fermentation, and combustion. A potential corollary of this theory of the role of air in chemical compounds was that fire might also be fixed in substances. Stahl saw phlogiston just that way, but he did not completely identify phlogiston (for him a reactant) with fire, which was for him an instrument. To Rouelle, by contrast, phlogiston and "matter of fire" were one and the same.

Rouelle's course in chemistry was catalytic for Lavoisier, who went on to study several of Rouelle's influences in the original, often commenting on them in his own memoirs. In his reading of *Elementa chemiae* he noted that Boerhaave "was not always in perfect agreement with himself about the combination and the fixation of air: sometimes he seems to deny that air could combine itself into bodies and contribute to the formation of their solid parts; sometimes he seems to adopt the opposite opinion." Underlying Boerhaave's uncertainty must be something to be ascertained. Here, as in so many departments of Lavoisier's career, the confusion of others provided him with an attractive point of entry.

AS EARLY AS 1766, when he had yet to win his provisional membership in the Academy of Sciences, Lavoisier had written, "Air is not a separate element. It is a compound." And still more assertively, "It is water turned into vapor, or, to be more precise, it results from the combination of water and the matter of fire." From the courses he had taken some years before,

Lavoisier was aware of Rouelle's equation of fire and phlogiston, but for some reason he chose not to use the latter term.

There is a limit to the degree of precision that can be attained in a formulation that involves a nonexistent substance such as matter of fire. Lavoisier, in these early days, was inclined to confuse the vaporization of water by heat with the chemical composition of atmospheric air. While speculating on the nature of the elements, he had been reading articles by J. T. Eller published in the *Memoirs* of the Berlin Academy. More or less in the tradition of Paracelsus and van Helmont, Eller spurned the four-element theory in favor of the "principles" of fire and water. Like van Helmont, Eller believed that water could be transmuted into earth, a point that Lavoisier would later take pains to disprove. But he thought Eller's notion that air might be accounted for by a combination of water with matter of fire to be worth consideration.

And here Lavoisier was thinking on similar lines as his eventual colleague in the French government, the Physiocrat Turgot, who coined the term *vaporize* and who reasoned (in his anonymous article "Expansibility" in Diderot's *Encyclopédie*) that steam was produced by a combination with heat (or matter of fire, or phlogiston). This reasoning was a way of accounting for the changes of state that a substance like water could be observed to undergo: from solid to liquid to gas. Turgot went so far as to postulate that *all* substances, at least in theory, could exist in all three of these states.

Still in 1766, Lavoisier sketched a note to the effect that air might be an expanded fluid—a liquid having undergone a change of state by combination with the matter of fire (also known as the "igneous fluid," the nonexistent substance that

Marat would pretend to make visible for the Academy examiners a few years later). In its state of expansion, air was elastic. But air could also be fixed in various substances, as Lavoisier knew from the work of Stephen Hales—mentioned both by Rouelle in his chemistry courses and by Eller in the articles Lavoisier studied circa 1766; a French version of Hales's *Vegetable Staticks*, translated by Buffon, had been available since 1735.

Fixed air lost its elasticity and was compressed into a much smaller space than air in the "expanded fluid" state. Hales had devised various experiments that captured the air released from such fixation in processes like fermentation, distillation, and even the respiration of small animals. The fixed air Hales measured in these experiments was usually carbon dioxide, though chemistry was not yet capable of analyzing it; Hales was capturing the bubbles from your beer.

Pondering the release of fixed air in Hales's experiments, Lavoisier recalled that certain effervescent reactions were known to produce a cooling effect. That phenomenon seemed consistent with a theory that heat (or phlogiston, or matter of fire) was absorbed into the vapors produced by effervescence. It was also known that the temperature of melting water and ice does not increase with the amount of heat applied to the melting process. Lavoisier took this to support the reasoning that heat (or phlogiston, or matter of fire) was fixed in the melting process by entering into the composition of water as it melted from ice—as air could be fixed in substances.

Then there were the spectacular experiments on the incineration of diamonds. By the spring of 1772, when he participated in these, Lavoisier was already completely committed to the axiom of conservation of matter—nothing is created, nothing is destroyed. Where, then, did the diamonds go? The fact that the diamonds could only vanish in the presence of air was intriguing, and clearly somehow related to the longer train of Lavoisier's thought during this period, though for the moment he could not explain it. The experimental results were insufficient to determine whether the diamonds "volatilized"— evaporated—or whether they "decrepitated" into fragments too small to be discerned by the experimental process used. Lavoisier designed a couple of more subtle experiments to settle this question, but these were never carried out; perhaps the supply of disposable diamonds was shut down.

ON AUGUST 19, 1772, Lavoisier made a presentation to the Academy of Sciences entitled "Memoir on Elementary Fire"; the written version of this lecture, drafted on August 8, was more prosaicly (and wordily) headed "Reflections on Experiments Which One Might Try With the Aid of the Burning Glass." The occasion for these notes was to list a series of experiments to be performed with the Tschirnhausen lenses, the apparatus used to incinerate diamonds in the Jardin de l'Infante.

Lavoisier began the paper by sketching the existing theoretical base, noting at the outset that "the theory of Stahl On phlogiston and the reduction of Metals" had been prevalent in Germany for a long time before being introduced into France. He credited the first French appearance of phlogiston theory to the 1723 publication of a "Course of chemistry Following the principles of Stahl and of Newton," then noted that Stahl's essential points were supported by experiments described in 1709 by a Frenchman, M. Geoffroy *l'ainé*—experiments that had also been conducted with a burning glass.

Geoffroy's conclusion was that "all metals or metallic substance are composed 1st. of a Vitrifiable earth, particular to each one of them. 2nd. of an oil or of an inflammable principle, the same which is found in Plants, in animals, in Charcoal and he observed that this Substance can Separate Itself from Metals, that one can take it out of them, and put it back into them at will, make it pass from one metal into another."

If Geoffroy's description looked rather opaque, Lavoisier knew how to illuminate it. "It is easy to see," his lecture went on, "that this System does not differ from that of Stahl except in that M. Geoffroy calls oily Matter or inflammable Substance that which M. Stahl names phlogiston; well, it must be confessed that even today we don't yet Know the nature of what we call phlogiston well enough to be able to declare anything very precise About Its nature."

Exactly there was the problem, as Lavoisier had constructed it—by penetrating the apparent differences between Stahl's terminology and Geoffroy's to pinpoint the difficulty neither had resolved. So far as Lavoisier was concerned, phlogiston would be the ultimate target of the program of experiments proposed.

The advantage of the Tschirnhausen lenses for this project, Lavoisier explained, was that the burning glass could concentrate intense heat on objects sealed in a vacuum, while "the fire which chemists are accustomed to employ cannot ignite or subsist in the Void. air is an agent necessary to Its conservation. the fire of the burning Glass offers a great advantage in this regard. it can penetrate the Receptacle of the pneumatic Machine And one can by Its Means do Calcinations and Combinations in the Void."

What follows is a list of experiments to be performed with

the Tschirnhausen lenses on metals, stones, crystals including but not limited to diamonds, and certain fluids—few of which, as Lavoisier noted, had ever been previously tried. Despite his emphasis at the outset, he does not mention phlogiston again in this paper. But (with the benefit of hindsight) his conclusion can be read as a veiled and cautious sketch of a competing theory.

The concluding section of Lavoisier's "Memoir on Elementary Fire" says nothing at all about fire as an element. The heading is "ON FIXED AIR, or rather, on the air Contained in Bodies." Here Lavoisier notes, "It seems Constant that air enters into the Composition of most Minerals, even Metals and in very great abundance. No chemist, however, has yet made air enter into the definition either of Metals or of any mineral body." Between these lines may be read a hint that Lavoisier himself might soon enough emerge as such a chemist. Though he does not say it unmistakably, the tendency of the concluding section is to discuss air as a component in minerals and metals *instead of* elementary fire, phlogiston, oily matter, or whatever anyone might want to call it. But rather than rushing to that conclusion, Lavoisier ended by remarking, with a certain coyness, that a great deal of work remained to be done.

AS EARLY AS 1762, the Scottish chemist Joseph Black had offered a theory of latent heat which explained the capacity of melting ice to absorb heat without increasing its temperature. Black's work, like Stahl's, was slow in becoming known in France (had Lavoisier known of Black's work with fixed air, he could not have honestly written in August 1772 that no chemist had ever incorporated air into the definition of minerals and metals). A few days after Lavoisier's lecture on elementary fire, news of Black's theory of latent heat suddenly penetrated the French Academy of Sciences, which reacted by delving a 1750 paper on a similar subject by Nollet from its archives, so as to stake a claim to French priority in this research. Such gestures of scientific national chauvinism were common enough; indeed, Lavoisier had done something similar in his own recent lecture, with wording that tended to confuse priority in phlogiston theory between the German Stahl and the Frenchman Geoffroy.

The presentation of Nollet's old paper, coupled with the news of Black's latent heat theory, seemed to strike Lavoisier's personal priority nerve. When the reading concluded, he rushed from the salon and swiftly returned with a document of his own for the Academy's secretary to initial. What it contained was a preliminary and fragmentary new theory of the elements.

IN THE CONCLUSION of Lavoisier's August 19 lecture, and in the midst of hinting at the intriguing possibilities of the idea that air entered into the composition of metals and minerals, he curbed his own enthusiasm by saying that for the moment, "We will not Follow these Views any further. they are the subject of a work already much advanced and even in part Drafted." Then, as if unable to contain himself completely, he went on to announce, "These views if followed to profundity might Lead to an interesting theory which we have even already sketched—," but there he broke off, more or less in midstride, to resume a description of the effervescence of metals under the heat of the burning glass.

The manuscript that he rushed to the Academy for the secretarial signature following the reading of Nollet's work seems to be the same draft or sketch to which he had referred in his own talk on August 19. The work is evidently incomplete and even somewhat garbled. Never published, it was obviously written under the frantic pressure of intellectual excitement, and it also appears to have somewhat mixed intentions—in some passages Lavoisier seems to be preparing his text for public presentation as a finished work, while in others he is frankly struggling, his pen as his instrument, with intractable aspects of the theory in progress, which he was so far unable to complete.

Most likely he was well aware of these difficulties, since he tried to finesse them at the very end: "I beg the public to pardon me if I have entered into somewhat extensive detail ~~but the main object is however to be clear when it~~ so as to lead it to my opinion every new idea demands a kind of preparation to be accepted and to make myself heard I am obliged myself to Guide the reader by the route which I Followed myself in my ideas. . . . Here are my ideas on the elements." The path that Lavoisier was tracing through his ideas at this point was by no means as straightforward as he wanted it to be; seventeen years later, when he published the definitive version of his views on the elements, any deviations from the logical high road had been erased. Given the disintegrating sentence structure and frequent cross-outs, one can picture Lavoisier scribbling his *faux* conclusion as he hastened from his laboratory back to the Academy, where a discussion of Nollet's paper was still going on. The secretarial signature appears immediately below.

Though it does not live up to the description "system of the elements," Lavoisier's rough draft does represent his first effort to organize everything he knew or dared to suppose about the elements into a comprehensive theory. In his haste he did not take time to lay out the axioms of his reasoning (or perhaps he had already begun to doubt them), but the text shows him still to be struggling with the problem of bringing Aristotle's four-element theory into accord with Stahl's phlogiston theory—a problem he had inherited most directly from Rouelle. He is trying to formulate an organized discussion of how fire, air, water, and earth enter into the composition of bodies—yet his progress is frequently derailed by phenomena still lacking an adequate explanation: evaporation, effervescence, the issues involved in latent heat, and the ability of substances to exist in two or three different states without changing their chemical composition.

Phlogiston or matter of fire was still a cornerstone of chemical orthodoxy, and Lavoisier's text tries to handle it as such—sometimes. "Matter of fire," he suggests, exists in "two different states in nature, 1st as combined with the other elements ~~with all the bodies As we observe it equally of air~~" and "As a Stagnant fluid which penetrates the pores of all objects and which puts Itself more or less in equilibrium in each of them and whose greater or lesser intensity produces the different degrees of heat"—a description of the notorious but nonexistent igneous fluid. At other times, though, Lavoisier doubles back on his own deployment of the phlogiston concept (most often when describing effervescence and other reactions producing a dramatic release of air from substances): "What we have just said relative to air we may equally say about phlogiston or matter of fire." Statements like these seemed to open the door to substituting air for phlogiston in the theory of the reactions being observed. On average, though, Lavoisier was more or less sticking with the view that "fire enters into the composition of all Bodies."

The last brief section of the 1772 "system of the elements" document has a new heading—"Reflections on Air"—and also a fresh date of August 1772. It reads as if Lavoisier had quickly reviewed his work as he rushed it to the Academy for signature and suddenly recognized that what it produced was not a conclusion but a question:

> But does air exist in Bodies How can This fluid susceptible of such a terrible expansion fix itself in a Solid and there occupy a space six hundred times less than it occupies in the atmosphere? How to conceive that that the same body can exist in two such different states?
>
> The Solution to This problem tends toward a Singular theory which I am going to try to make understood that is the air that we breathe is not at all a simple being It is a particular fluid combined with the matter of fire—

But here Lavoisier, now completely out of time, broke off without quite finishing this thought, and began praying for the reader to pardon the crooked path of his thinking, and so on. And yet, he *had* arrived (at a breakneck pace on a winding road) at a radical implication. Suddenly, air was no longer an element; it had become a compound.

IN THE OPENING passage of the August 1772 "system of the elements" manuscript, Lavoisier had admitted that for the moment "we don't have any new experiments to offer here but we have worked to assemble those which exist in order to draw Conclusions from them." Sometime during this same period—the summer or fall of 1772—Lavoisier made some undated but pointed notes on a work called *Digressions académiques* by the chemist Guyton de Morveau (later to become one of Lavoisier's staunchest allies in the push for the acceptance of the new chemistry and the new terminology that went with it). Guyton had performed a series of quite rigorous experiments which established definitively that various metals did gain weight when calcined. He also did another group of experiments which showed that calcination could not occur in sealed vessels deprived of air (another point of sharp interest for Lavoisier). However, Guyton furnished an explanation of the augmentation of weight by using a variation on the version of phlogiston theory preferred by his older, recently deceased colleague Jean-Pierre Chardenon. Guyton's paper "Dissertation on phlogiston considered as a weighty body and with regard to the changes in weight it produces on the bodies to which it unites" was published in 1772 as the first chapter of his *Digressions académiques*; there Guyton explained the weight gain of calces, which had, according to Stahl's chemistry, lost phlogiston in the process of calcination, by asserting that phlogiston is lighter than air.

What struck Lavoisier was that Guyton had fully established the reality of weight gain in metallic calces by a set of experiments fenced by the kind of strict and thorough measurements that he himself preferred. Guyton's work inspired Lavoisier to jot an efficient summary, "On the matter of fire":

All metals exposed to fire and calcined very noticeably increase in weight.

The ancient authors claimed that one combined fire with these bodies during calcination and that it was to the addition of this weighty Substance that one owed the augmentation of weight.

Stahl claimed that calcination removes the matter of fire from bodies that one calcined but he and his partisans fell into a labyrinth of difficulties how to conceive in effect that one augments the weight of a body by taking part of its substance away from it.

Whatever the explanation may be, the fact is no less constant. All metals gain weight by calcination. M. De Morvaux demonstrates it completely in His academic digressions, page 72 to 88.

Here, as in the hasty effort to tack some sort of conclusion onto his "system of elements" manuscript, Lavoisier had managed to focus on a crucial question. In the long run, the two questions would turn out to have a single answer.

THE TSCHIRNHAUSEN LENS apparatus was an exciting device to the eighteenth century; historian Arthur Donovan compares it to a particle accelerator in the hands of twentieth-century researchers: "a machine capable of blasting apart substances previously thought to be immutable." The lens was a very powerful weapon for attacking substances supposed to be elements, to determine if they were irreducible or not. The collaborative program of experiments with the great burning glass that Lavoisier had proposed in August 1772 continued in October, but since he was still a very junior member of the team, his own most interesting ideas for experiments were not given much priority until the elder scientists had finished their own work. In mid-October, Lavoisier was finally able to use the great burning lenses for his own ends.

In the 1680s Robert Boyle had published exact specifications for the preparation of phosphorus from evaporated urine; Lavoisier preferred to purchase an ounce of the material from another French scientist, Pierre François Mitouard, for the noteworthy price of forty-five livres, or about eighteen hundred U.S. dollars in the twentieth century. On October 20, 1772, he used the burning glass to ignite eight grains of the phosphorus under a bell jar; after the combustion a quantity of phosphoric acid condensed—of a much greater weight than that of the phosphorus burned. Lavoisier attributed the weight gain to the fixation of air in the phosphorus. Not long after, he calcined sulfur under a bell jar and found, again, that the weight of the sulfuric acid resulting was greater than that of the sulfur he had heated.

At the very end of his presentation of experiments to be undertaken with the burning glass—the subsection entitled "ON FIXED AIR"—Lavoisier concludes with this suggestion: "It Would be much to be desired to apply the burning Glass to the apparatus of M. Halles so as to measure the quantity of air produced or absorbed in each operation, but one fears that the difficulties which this kind of experiment presents might be insurmountable for the burning glass." These fears would be proved, at least in part, though Lavoisier did obtain strong enough results to bolster his emerging theory.

The Englishman Hales, to facilitate his study of fixed air, had invented a device he called the *pneumatic trough*, in which gases released in a reaction were channeled from the reaction vessel into an inverted dome filled with water—a reliable system for capturing and retaining airs. A variation on this scheme involved placing a pedestal in a basin of water. The experimental object (sometimes a small animal, sometimes a burning substance) was perched on the pedestal and covered by a bell jar with its open end submerged in the water. By calibrating changes in water level, the experimenter could measure the gain or loss of air or gases affected by whatever reaction takes place on the pedestal.

Lavoisier's adaptation of Hales's pedestal apparatus was nicely drawn by his wife as an illustration for his *Opuscules physiques et chymiques*, in which the results of these experiments were recorded. In his version, he set a porcelain crucible on a crystal pedestal, covered it with a bell jar, and controlled the water level with a siphon. A layer of oil on the surface of the water under the jar prevented any gases released from dissolving. In October of 1772, he put a lead oxide called *minium** into his crucible with a small amount of charcoal and heated it through the bell jar with the burning glass, whose beam was narrowly focused on the contents of the crucible. This method, common for smelting metals from oxide ores, produced a significant release of gas of some description, or elastic fluid. Although (uncharacteristically) Lavoisier did not make an exact measurement of the gas, he recorded "a volume at least a thousand times greater" than that of the lead oxide used.

What actually happened in this experiment, though

---

* Pb_3O_4. In discussing this experiment, Lavoisier sometimes says that litharge (PbO) was reduced, and sometimes minium; the latter seems more likely. See Henry Guerlac, *Lavoisier—The Crucial Year: The Background and Origins of His First Experiments on Combustion in 1772* (Ithaca: Cornell University Press, 1961), p. 160.

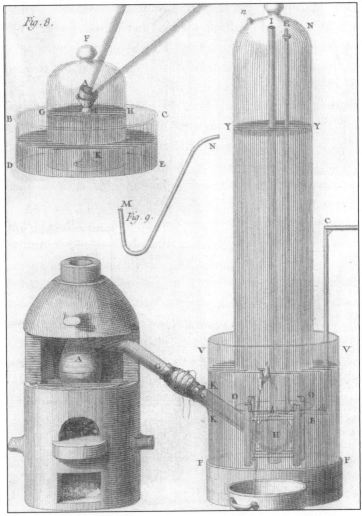

Lavoisier's apparatus for his experiments on minium, as drawn by Madame Lavoisier.

Lavoisier did not yet know it, was that the oxygen released in the reduction of lead oxide combined with carbon in the charcoal to form a large volume of carbon dioxide (fixed air). For his immediate purposes, the minium experiment demonstrated and confirmed an already known fact: the reduction of lead oxide released a gas. It also complemented the syntheses he had performed by combining air, phosphorus, and sulfur with an analysis where air was subtracted from a calx. Lavoisier's notion that air was fixed in the process of calcination (and released in the reduction of calces) was now supported at both ends.

Lavoisier was satisfied (or at least willing to claim) that these results had "Completely Confirmed my Conjectures." The experiments on phosphorus, sulfur, and minium became the basis for the sealed note he deposited at the academy on November 1, 1772. The truth was that even though Lavoisier was convinced his discovery was "one of the most interesting that has been made since Stahl," he still could not say precisely *what* had been discovered—nor did the discovery yet fit comfortably into his evolving theory. Not until February of the following year would he feel confident enough to declare, in that famous lab notebook entry, that he was going to bring about "a revolution in physics and chemistry."

THE NEXT FEW months were hectic. As much as any modern scientist on the trail of a patent, a cure, or the glory of a major prize, Lavoisier felt the pressure of competition. The *pli cacheté* reveals his anxiety that he might accidentally give away his idea to one of his French colleagues, and, in fact, there were French chemists—Guyton de Morveau and Pierre-Joseph Macquer, among others—busy at work in the promising field. Of course, since the gas had not yet been properly defined, none of the competitors knew that oxygen was what they were trying to discover, but they did all understand that something major was to be learned from pneumatic chemistry—the chemistry of air.

Hindsight shows that Lavoisier's strongest competition came from outside France, and he was certainly aware of that in 1772; he had several reasons to be thoroughly conscious of the progress of the British pneumatic chemists, not only Hales and Black, but also, increasingly, a man named Joseph Priestley. A clergyman and doctor of theology, Priestley was a dissenter from the doctrine of the Church of England; he was denounced for his rejection of the idea of the Holy Trinity and so excluded not only from the official Church but also from Oxford and Cambridge Universities. Inclined to see himself as a natural philosopher of the kind whose thinking embraced both religion and science, Priestley had his interest in the latter focused when he began preparing to teach the sciences in one of the academies founded by other dissenters shut out of the official educational system. Priestley was a maverick personality in almost every way, and he did not share Lavoisier's systematic turn of mind, but (in the Baconian tradition) he was an enthusiastic collector of curious facts, and some of his findings were novel enough to attract the attention of contemporaries as august as Benjamin Franklin, even before the French took notice.

During his peregrinations, Priestley spent a year next door to a brewery in the city of Leeds, where he noticed that the beer-making process emitted copious quantities of the fixed air eventually to be defined as carbon dioxide. He did not imagine that he had discovered this gas—Joseph Black, in 1757, had shown that respiration produced a nonbreathable fixed air—but he devoted some energy to exploring its properties. Priestley's work demonstrated that this fixed air did not support combustion or respiration, but that its breathability and ability to nourish a flame could be restored by plants growing in it. He also developed a method of carbonating water, which he published in June of 1772 as a pamphlet titled *Directions for Impregnating Water with Fixed Air*.

Priestley was instigated in the latter project by two British medical doctors, John Pringle and David MacBride, who foresaw medical uses for carbon dioxide and even believed (erroneously as it turned out) that fixed air might prevent scurvy among maritime crews. Scurvy had such huge implications for both mercantile and naval power during the eighteenth century that this unjustified hope for a cure suddenly endowed anything connected to fixed air with the aura of a military-industrial secret.

For that reason, the Portuguese monk Jean-Hyacinthe Magellan, who reported many developments in British chemistry to the French chemists, is sometimes referred to as a spy. The designation seems exaggerated; the discoveries of Priestley, Black, and their colleagues were not kept secret in England but were promulgated both in published texts and in lectures before the Royal Society of London, an analogue to the French Academy. Still, Magellan was quick and vigorous in reporting British advances to France, where his contact was a nobleman, amateur chemist, and honorary member of the Academy of Sciences, Trudaine de Montigny.

Trudaine's attitude to Lavoisier combined the qualities of

patron and fan. He knew that Lavoisier was much closer to the cutting edge of chemical research than he himself was likely to come, but he was not above advising Lavoisier from time to time, and sometimes his advice was good. Priestley had presented his research on fixed air in a lecture to the Royal Society in March of 1772, then published his work in *Directions for Impregnating Water with Fixed Air*. Trudaine received this pamphlet from Magellan in July of 1772, along with Magellan's report on Priestley's work, and within a week had forwarded it to Lavoisier, with a letter exhorting him to translate the text into French and to repeat and confirm Priestley's experiments. "I know your exactitude in the details of physics and chemistry," Trudaine wrote, "and I know that I am doing you a favor by putting you in the way of doing something useful." On July 18, Lavoisier duly presented Magellan's description of Priestley's work to the Academy; a French translation of the carbonation pamphlet appeared the next month.

These events took place several months *before* Lavoisier deposited his sealed note on weight gain in calcination at the Academy, and during that time he must have been working with the knowledge that the British pneumatic chemists were abreast, if not in fact ahead of his own research. What he could not know was whether or not one of the British chemists might be at work on a theory as comprehensive as the one he was trying to bring to completion. But the aspect of international competition contributed to his sense of urgency.

"The more the facts are extraordinary," Lavoisier wrote, "the further they are removed from received and accredited ideas, the more important it is to confirm them by repeated experiments and in a manner such as to leave no doubt." The decided, epigrammatic style of this statement is a far cry from Lavoisier's hastily scribbled drafts and notes of late 1772. But the *pli cacheté* of November 1 required him to make good on it—to confirm his conjecture by experiments transforming doubt into certainty.

Perhaps the sealed note overreached what he would be able to prove. One way or another, the study of fixed air (which Trudaine de Montigny had urged) was likely to decide the question. When Lavoisier assembled the information on fixed air available from both his own experiments and those of others, it seemed to be rather slippery stuff: sometimes it extinguished flame and killed animals; at other times it brightened flame and seemed better for breathing than ordinary air. The solution to this contradiction was that the current usage of fixed air included both carbon dioxide and oxygen, but Lavoisier was still a good distance from recognizing the difference between them.

Four months later, with few of his conjectures certainly confirmed by experiment, he was still optimistic enough to open that famous laboratory register with the declaration that he would enact a revolution in physics and chemistry. Then he laid out his program for doing so:

> This manner of viewing my objective has caused me to perceive the necessity first to repeat and then to multiply the experiments which absorb air so that, knowing the origin of that substance I can follow its effects in all the different combinations.
>
> The operations by which one can arrive at fixing the air are vegetation, the respiration of animals, combus-

tion under certain circumstances, and finally certain
chemical combinations. It is by these experiments I
believe I must begin.

When he was finished, he would have a much clearer
understanding of the nature of fixed air and its relationship to
the air we breathe.

Clear and systematic as this program was, it got swamped
in technical difficulties at the very start. Earlier, in his presentation to the Academy in August of 1772, Lavoisier had fretted
that the Hales apparatus might be difficult to combine with
the Tschirnhausen lens, and in his first attempts the bell jars
did, in fact, crack under the intense heat of the burning glass,
preventing measurement of the gases released. On February
22, 1773, he tried to calcine lead (an operation that fixed air)
by heating a retort* in a conventional furnace, but the experiment failed when the retort cracked. He had sketched designs
for several new pieces of equipment, including a "machine to
test the effects of air on animals," but his workmen could not
produce them on time, and time was of the essence: Lavoisier
badly wanted definitive results to present to a public meeting
of the Academy coming up in the third week of April.

Despairing of the new gear he had ordered, he improvised
a version of the Hales apparatus using ordinary glass jars, a
hand-washing basin, and a crystal pedestal meant for the dec-
orative display of fruit. Applying the burning glass to this rig,
he succeeded in starting a calcination of lead on March 29, but
was puzzled when the process stopped on the surface of the

---

* A glass vessel with a long snout used for distillation, the retort was the
symbolic instrument of both alchemy and early chemistry.

metal; he imagined various explanations for this result but had no way to determine which of them was correct. Other experiments during these weeks, including efforts to calcine tin and detonate niter with sulfur, had similarly frustrating and ambiguous results.

By April, Lavoisier had made far less progress than he had hoped toward the experimental confirmation of his ideas that he had promised himself—and that the sealed note of the previous fall implicitly predicted. Nevertheless, he pressed on with the presentation of the theory at the Academy's meeting on April 21.

"Present circumstances do not permit me to give here the detail of my experiments," he announced—which, considering that none of them had yet worked out as expected, was something of an understatement. Early drafts alluded more plainly to the discrepancy between his interpretations and any actual results, but by the time of his presentation to the academy, Lavoisier had revised those sentences completely out of existence. Though the language he used obscured the point, the experiments he described in detail were really "thought experiments" at this stage, rather than anything that had yet been realized on the material plane:

> If [a big "if," under the experimental circumstances] instead of doing these experiments in open air one does them in a portion of air closed under a glass bell reversed in a . . . trough and if one intercepts the communication with the air of the atmosphere . . . to the measure that these metals reduce themselves to calx the volume of air diminishes and augmentation of weight of the metal is found to be more or less equal to the quan-

tity of air absorbed. If by means of a burning glass or by some other procedure of which I will give the details, one arrives at the reduction of these metals—that is to make them pass from the calx state to that of metal— soon enough they render up all the air they have absorbed and at the same time lose again the augmentation of weight they have acquired.

Though still a hypothesis disguised as a result, this "outcome" was tremendously significant, permitting Lavoisier to go forward toward definitive conclusions about the calcification of metals and reduction of calces: "It obviously results from these experiments, 1st that a metallic calx is nothing other than the metal itself combined with fixed air, 2nd that the metallic reduction consists only of the disengagement of the air from metallic calces. 3rd that the metals owe the weight gain* to the fixed air contained in the atmosphere."

From this position, Lavoisier could fire his first open shot across the bows of Stahl's phlogiston-based chemistry: "This theory is destructive to that of Stahl adopted by almost all chemists, and that circumstance would be suitable to put me on guard; however, I could not refuse the evidence; above all decisive experiments have assured me that it is possible to reduce almost all metals without the addition of phlogiston." The word *decisive* is rather extreme, given the real state of Lavoisier's experimental program at this time; a correction in the margin suggests that he may have put this point more cautiously during his actual performance before the Academy. Still, his conclusion was very bold: "I have even come to the

---

* Observed in calcification.

point of doubting if what Stahl calls phlogiston exists, at least in the sense that he gives to that word, and it seems to me that in every case one could substitute the name of matter of fire, of light, and of heat." Though short of a full-fledged revolution, this statement was certainly a fervent gesture of rebellion.

Two weeks following the April presentation of his theory, on May 5, 1773, Lavoisier requested that the sealed note of the previous November be opened in the presence of his colleagues in the Academy, a ceremony that duly took place. Why he felt the need for it at this point is uncertain. The secretary of the Academy noted that "the author has asked the present mention to preserve for himself the date." So the priority issue must have been his motive, and yet it does not appear that any other theoretician was obviously breathing down his neck. He may have believed that once he had stated his theory so clearly, other chemists would begin to experiment along similar lines (and perhaps with better success than he had so far obtained). Or perhaps he simply recognized that once his theory had been unveiled, there was no particular reason to keep the November note under seal.

By bracketing his April lecture with the deposit of the November note and its subsequent unsealing, Lavoisier had staked his claim as well as he could—and he was the only one who knew how far he still had to go to justify it. Observers, at first, were skeptical. Benjamin Franklin, whose work with electricity during the previous decade had made him an international scientific celebrity, looked askance at Lavoisier's propositions at first, writing to Jean-Baptiste LeRoy of the French Academy, "I should like to hear how M. Lavoisier's doctrine supports itself, as I suppose it will be controverted." Lavoisier may or may not have got word of this aspersion, but he certainly knew that Franklin's opinion counted enormously in the scientific community, perhaps more than that of any other individual, and in the coming years both he and his wife would make every effort to win it to their cause.

HE HAD STUCK his neck a long way out, as people tend to do when they *know* they are right—availability of proof notwithstanding. Often such people are proved wrong, but Lavoisier's intuition was sound, and in the summer and fall of 1773 he gathered experimental evidence that really did support it. Premature as it was, the April announcement of his theory had bought him some breathing room. And finally, the various pieces of new equipment he had designed were delivered by their fabricators. A new and much more reliable version of the Hales device, now connected to a furnace and retort, both of which had been proofed against cracking or leaking, was put into play in the summer of 1773. Lavoisier modified an existing instrument called an *areometer* to give him more precise measurements of the density of fluids. And he constantly sought to improve the precision of his balance scale—sparing no expense. Eventually his laboratory would boast a balance accurate to one part in four hundred thousand.

The "thought experiments" in the calcination and reduction of metals on which Lavoisier had based the theory he presented on April 21 were also "balance sheet" experiments, founded on the principle of conservation of mass and dependent on the exact measurement of weights and volumes of materials used, before and after the processes. Throughout the summer of 1773, Lavoisier worked to bring his experiments from the realm of thought to the material plane, and by the end of the summer the balance sheets he scrupulously kept had finally begun to reconcile.

Theoretically, however, he found himself balked. In April of 1773 he had felt confident that the fixed air contained in bodies, which he as often called "elastic fluid," would account for a whole range of chemical reactions that in Stahl's chemistry were explained by phlogiston. However, conclusive evidence was not forthcoming. Lavoisier imagined his elastic fluid as a subcomponent of atmospheric air (which was true) but also he persistently misconceived it as a single substance (as it had been defined by others before him). In reality, his experiments were sometimes producing carbon dioxide, sometimes oxygen or other gases, and Lavoisier could not reconcile their different behaviors with the elegant simplicity of his theory in its current state.

In response to these difficulties, he continued to refine his experimental procedures while somewhat retrenching his theoretical position. It seems likely, too, that he may have been counseled by elder members of the Academy to be more cautious on the theoretical front. Phlogiston was welded into the foundation of the current chemistry, and the elder scientists would not easily be detached from it.

Nevertheless, in July of 1773 Lavoisier began to present new papers to the Academy, based on his research program still in progress, and he found some encouragement to collect them into a book. By August he had a draft of the work in hand. Earlier that month, Trudaine de Montigny (ever perspicacious for an amateur) recommended to him some refined experiments in the calcination of metals that looked likely to yield a firmer understanding of the nature of fixed air, but Lavoisier, in haste to complete his treatise, did not take time to try them right away. He did, however, manage to work all the experiments he had conducted since February into the text he was writing.

On August 7, 1773, he handed in his draft of *Opuscules physiques et chymiques* to be reviewed by the Academy for publication. One of the two scientists assigned to examine it was Jean-Baptiste LeRoy, the recipient of the skeptical letter from Benjamin Franklin. The next month, on September 25, LeRoy met with Trudaine, Macquer, and Louis-Claude Cadet to verify the most important experiments Lavoisier's treatise discussed. The encounter seems to have been quite a collegial one, for the group spent several days in Lavoisier's laboratory, often changing the tactics of experiments Lavoisier had previously performed. With Trudaine on the scene, some of the refinements he had suggested were now carried out. The scientists collected and measured the air released in the reduction of minium, and established that it extinguished candles, killed sparrows, and precipitated limewater* (all effects known to be associated with the type of fixed air that was actually carbon dioxide). A very simple test established that "air of respiration" (that is, the carbon dioxide that is a by-product of breathing) precipitated limewater, while ordinary atmospheric air did not.

Invigorated by the collaboration, Lavoisier kept performing calcination experiments into October, finally approaching a conclusion that "all of the air that we respire is not suitable to enter into combination with metallic calxes; but that there exists in the atmosphere a particular elastic fluid that is found

---

* Carbon dioxide dissolved in limewater $\{Ca(OH)_2(aq)\}$ precipitates carbonate ion $\{CaCO_3\}$.

mixed with the air, and *that it is at the moment at which the quantity of this fluid contained under the jar is exhausted that the calcination can no longer take place.*"

Here was a promising explanation for the stalled calcinations of lead that had frustrated him months before. Lavoisier was describing the combination of oxygen extracted from atmospheric air into the metallic calx, though he could not yet define it.

The academy commissioners delivered their evaluation of Lavoisier's *Opuscules physiques et chymiques* in December of 1773, noting that he had "submitted all his results to measurement, to calculation and the balance: a rigorous method which, fortunately for the advancement of chemistry, begins to be indispensable to the practice of that science." They also made a point of complimenting his restraint: "M. Lavoisier, so far from delivering himself too much to his convictions, contents himself to propose them once and in a few words, with all the reserve that characterizes enlightened and judicious physicists."

Indeed, Lavoisier had pulled in his horns a good deal since the previous April. The Academy's approval process had brought him into a more faithful conformity to his own contention that the more facts seem to run against received ideas and what is previously known, the more they must be confirmed by incontrovertible evidence. If he had overreached himself the preceding spring, in *Opuscules* he took care to stay within the bounds of what he could certainly prove. Even the diminutive title suggests the modesty of the ambition there expressed. These brief essays were only a harbinger of the grand opus Lavoisier intended down the road. However, he was quite energetic in getting the work out, both at home and abroad. He sent his *Opuscules* to the Royal Society of London, where he knew it would reach Joseph Priestley, and to the Royal Society of Edinburgh, with an extra copy especially for Joseph Black.

The accompanying letters paid respectful homage to the work of Lavoisier's predecessors on fixed air. The packages, taken as a whole, defined the direction of new research. The publication of *Opuscules* did not so much fire the starting gun in the race for oxygen as to lay out the track for it—and as yet no one could see the finish line.

The *Opuscules* volume was divided into two parts, the second of which detailed Lavoisier's experiments, with illustrations of the apparatus and procedures painstakingly drawn by Madame Lavoisier. The results were sufficient to consolidate his hypothesis that "some sort of elastic fluid contained in air" was the substance fixed in bodies, without advancing much beyond it—he could not advance much further while remaining on firm experimental ground. The first half, nearly two hundred pages long, was Lavoisier's history of all the work that had been done in fixed air before him. He had been reviewing this record, and writing it up, during the same extremely busy months of 1773 when he had carried out as much of his experimental program as he could complete.

Though Lavoisier was familiar with Hales's analysis of air, the advances made by Priestley and Black were quite new to him, and he knew that it was important to master them. Indeed, many of the experiments in his 1773 program were modeled on experiments Black had performed earlier to demonstrate the behavior of fixed air in transit from alkalis to calcareous earths and in combination with acids. Priestley's achievements came in for a still more detailed treatment in

Opuscules's opening "*Précis historique.*" Lavoisier's summary had enough historical depth that he also discussed the ideas of Paracelsus and van Helmont.

Paracelsus had noted an elastic fluid that escaped from combustion and fermentation, and named it "sylvan spirit." Following Paracelsian traces, as he often did, van Helmont had coined the term *gas* to describe it, and learned that some gases burned while others did not. Robert Boyle recognized that the gases constituted a group of substances as important as solids or liquids, and used an early version of the pneumatic trough to collect them. Without knowing what it was, he captured a quantity of hydrogen from the action of sulfuric acid on nails. John Mayow, one of Boyle's students, released oxygen by heating potassium nitrate in a vacuum in 1674—a full century before anyone had a context for defining it.

IN 1774 THAT context was rapidly coming together—and other chemists besides Lavoisier sensed it taking shape. Pierre Bayen, a military pharmacist, was accustomed to make different mercury oxides (calces) for practical use. Having observed that his mercury oxides lost weight while releasing a gas, he used the fixation theory of Lavoisier's *Opuscules* to explain it, and went so far as to say that this phenomenon challenged the whole idea of phlogiston. Phlogiston was thought to transfer to metal from charcoal in the conventional smelting of metals; Bayen learned that mercury oxide could be reduced without charcoal, which seemed to further invalidate the phlogiston concept.

A mild controversy broke out between two academic chemists, Antoine Baumé and Cadet, with the former maintaining that *mercurius calcinatus* could not possibly be reduced to mercury without the phlogiston derived from charcoal and the latter insisting that it could be. Lavoisier was one of a committee appointed to settle the question by experiment. The irritated Baumé boycotted the event. Cadet was proved right. The gas released in the reduction of *mercurius calcinatus without* charcoal was pure oxygen, but neither Lavoisier nor anyone else present noticed its difference from the fixed air or elastic fluid produced by other reactions.

Across the Channel, in August of 1774, Joseph Priestley took note that the gas collected from a reduction of *mercurius calcinatus* had an opposite quality to that of fixed air—"A candle burned in this air with a remarkably vigorous flame. . . . I was utterly at a loss how to account for it." Mercury oxide could be produced somewhat more easily than by calcination proper: the alternative method involved dissolving mercury in nitric acid and then reducing it to salt via techniques that sometimes introduced impurities. Therefore, when Priestley repeated his mystifying experiment several times with the same result, he began to question the quality of his sample. When he had the opportunity to visit Paris in October of 1774, he bought an ounce of *mercurius calcinatus* from Cadet "of the genuineness of which, there could not be any suspicion." The enterprising Magellan introduced them.

Priestley was fêted by the French chemists, and dined at *chez* Lavoisier. Six years later he recalled (rather pointedly) having mentioned his odd discovery to the others present: "saying that it was a kind of air in which a candle burned much better than in common air, but that I had not yet given it a name. At this, all the company, Mr. and Mrs. Lavoisier included, expressed great surprise." Lavoisier was doubtless kicking himself for failing to investigate the air produced from

Joseph Priestley

the same experiment he had recently performed to settle the dispute between Baumé and Cadet.

Priestley returned to England and went to work on the *mercurius calcinatus* he had obtained from Cadet. Animal experiments produced an unusual result—Priestley discovered that a mouse survived in his new gas for twice the time it could in the same volume of atmospheric air. Encouraged, Priestley inhaled some of the gas himself through a siphon. "I fancied that my breast felt peculiarly light and easy for some time afterwards. Who can tell but that, in time, this pure air may become a fashionable article in luxury. Hitherto, only two mice and myself have had the privilege of breathing it." Priestley was breathing pure oxygen, but he decided to call it "dephlogisticated air," reasoning that a loss of phlogiston made it superior to ordinary air. Theory was not his strongest suit, and he failed to consider that a loss of phlogiston should not have made the new gas more hospitable to combustion than atmospheric air, though he had noticed that it was.

A few years before either Priestley or Lavoisier did so, the Swede Carl Wilhelm Scheele had isolated oxygen from several different oxides. Scheele worked out of a Stockholm pharmacy nowhere so well or expensively furnished as Lavoisier's laboratory at the Arsenal; nevertheless his results were interesting. He obtained the new gas by heating materials like manganese oxide, silver carbonate, and potassium nitrate, but his improvised equipment limited what he could achieve. Nonetheless, he was a couple of years ahead of either Lavoisier or Priestley in isolating oxygen (in "an empty air bag") from the reduction of *mercurius calcinatus*. Scheele chose to call the new gas "fire air," and though he accurately identified its properties, he explained them in terms of phlogiston.

Lavoisier had some awareness of Scheele's work (despite distance and the language difference) via a correspondence between the French chemist Macquer and the Swedish professor Torbern Olof Bergman, who bought his chemistry supplies in Scheele's pharmacy. In April of 1774, Lavoisier shipped two copies of *Opuscules* to the Academy of Science in Stockholm with a note directing one of them to Scheele, with his compliments. On September 30, 1774, Scheele responded by offering Lavoisier his own research program, as it were, on a platter, though in the form of a brief handwritten letter rather than a bound and illustrated treatise. Perhaps it was fatiguing for Scheele to write in French; the language barrier

was a general impediment to scientific discoveries making
their way out of Sweden into the rest of Europe. However, he
managed to express himself quite clearly.

Monsieur,

*I have received from Mr. Secretary Wargentin a book
which he said you had had the goodness to send me.
Although I do not have the honor to be known by you, I
take the liberty of thanking you most humbly. I desire
nothing more ardently than to show you my gratitude. For
a long time I have wished to be able to read a collection of
all the experiments that have been done in England and in
Germany on all sorts of air. You have not only satisfied that
wish but also, with your new experiments, have given
Savants the most beautiful opportunities to better examine
fire and the calcination of metals in the future. During several years I have done experiments with several kinds of
air, and I have also spent a lot of time in discovering the
singular qualities of fire, but I have never been able to com-
pose ordinary air from fixed air: I have several times tried,
according to the opinion of Mr. Priestley, [to do it] with a
mixture of iron filings, of sulfur and water, but it has never
succeeded because the fixed air has always united itself to
the iron and made it soluble in the water. Perhaps you
don't know any way to do it either. Because I don't have a
large burning glass, I beg you to make a try with yours, in
this manner: Dissolve some silver in nitrous acid and precipitate it with tartar, wash this precipitate, dry it, and
reduce it with the burning glass in your machine fig. 8,**

*The reference is to the illustration of Lavoisier's adaptation of the Hales
apparatus, drawn by Madame Lavoisier for *Opuscules.*

*but because the air in this glass bell is such that animals die
in it and a part of the fixed air separates from the silver in
this operation, one must put a little quicklime in the water
where one has put the bell, so that this fixed air will join
itself more quickly to the calx. It's by this means, I hope,
that you will see how much air is produced during this
reduction, and if a lit candle can sustain its flame and if
animals can live in it. I will be infinitely obliged to you if
you let me know the result of this experiment. I have the
honor to be always and with much esteem*

> *your very humble servant*
> *C.W. Scheele*

With this, Scheele had furnished Lavoisier with an excellent
procedure for isolating oxygen, which the Swede himself, with
his ill-equipped lab, could not execute. Here was the first dramatic example of how the technological firepower that
Lavoisier took such pains to accrue could put him lengths
ahead in a research race. Lavoisier never answered Scheele's
letter (probably he wished he had never received it). Nor is
there any indication that he ever performed the experiment
Scheele had described. Quite likely, he realized that if he had
done so, he would have made an important discovery by following someone else's directions.

Instead, Lavoisier returned to the experiments with *mercurius calcinatus*. Here Priestley had a claim to priority, which
he would later assert with some vigor. And a French observer
of pneumatic chemistry on both sides of the Channel,
Edmond C. Genet, challenged Lavoisier about the resemblance of his work to Priestley's. Lavoisier laughed (just how
heartily has not been recorded) and replied, with both wit and

accuracy, "My friend, you know that those who start the hare do not always catch it."

Lavoisier may have felt safer with the *mercurius calcinatus* experiments than with the one suggested by Scheele, because after all he had observed such mercury experiments before he ever heard Priestley say anything on the subject, while adjudicating the argument between Baumé and Cadet. What Priestley had disclosed at the famous dinner (yes, it *is* difficult to avoid letting such matters slip among friends) was that the gas produced in the absence of charcoal was something different from the usual fixed air.

In any case, Lavoisier resumed experiments with calcined mercury in November of 1774, and by the end of March 1775, his results struck him as significant enough that he sealed them and deposited the envelope with the Academy of Sciences. The definitive *mercurius calcinatus* experiments finally drew a clear distinction between carbon dioxide and oxygen according to their qualities, though neither yet had been assigned these names. The gas produced by reducing *mercurius calcinatus* with charcoal was shown to dissolve in water, precipitate limewater, extinguish candle flames, and suffocate mice and birds; it was carbon dioxide—the fixed air described by Black, Priestley, and others. When *mercurius calcinatus* was reduced in the absence of charcoal, the resulting gas supported flame and animal respiration, but did not combine with water or precipitate limewater; it was Priestley's dephlogisticated air, or Scheele's fire air.

Lavoisier did not yet know what to call it, but he liked it: "We twice tried . . . the experiment of the candle," he wrote. "It is charming; the flame is much larger, much clearer and much more beautiful than in common air, but without color other than that of ordinary flame." Sometimes he referred to the new gas as "eminently breathable air." When the Academy reopened for Easter, on April 26, he was there to announce that "the principle which unites itself to metals during their calcination and which increases their weight and which constitutes them into the state of calx, is neither one of the constituent parts of air, nor a particular acid spread through the atmosphere; *it is air itself, entire, without alteration, without decomposition.*" To that extent, Lavoisier was still getting it wrong; he had not yet grasped that his "eminently breathable air" *was* a "constituent part" of atmospheric air, and not the whole thing. But during the same talk he also said that the new gas was "purer and more breathable . . . than the air of the atmosphere, and more suitable to sustain ignition and the burning of bodies." And he was describing the new gas in a way that contributed half of the phrase that would finally announce the discovery of oxygen: "*le principe* _____." It still remained to fill in the blank with a single efficient term.

SUFFICE IT TO say that the discovery of oxygen was never as clearly or brilliantly delineated as, say, lightning striking a key on a kite. Lavoisier was an ambitious young man, willing enough to clamber over his colleagues and competitors however he could to reach the goal. In 1775 he was surrounded by disgruntled competitors, not only Priestley (who had begun to grumble) but also Bayen, who in February isolated both oxygen and carbon dioxide in *mercurius calcinatus* reductions, but was far less astute than Lavoisier in discerning the difference between them. Annoyed that Lavoisier had aspired all the credit for the interpretation of these experiments, Bayen found and had published a 150-year-old book by Jean Rey, which remarkably predicted that the weight gain of metals in calcination was caused by the fixation of air. This maneuver sparked a debate that went on for some years, though in the end Lavoisier's claim to be first among French chemists in the discovery of oxygen was not seriously challenged.

Carl Wilhelm Scheele was a less ambitious, more modest personality than most of these others. Perhaps the humility he described in his revealing letter to Lavoisier was not merely rhetorical, or perhaps he felt that his position was too weak to support a claim to credit that might well have been his due. Scheele had isolated oxygen—fire air in his term—as early as 1771 and had completely described its properties well before Lavoisier or anyone else, but he did not publish his *Treatise on Fire and Air* until 1777, and even then he made no claim to priority. By then, Lavoisier's claim had grown too strong.

Lavoisier's talk to the French Academy in April of 1775 was published the next month as "On the nature of the principle which combines with metals during calcinations and increases their weight"; it prompted a pointed reaction from Priestley:

> After I left Paris, where I procured the *mercurius calcina-tus* above mentioned, and had spoken of the experiments I had made, and that I intended to make with it, he (Lavoisier) began his experiments on the same substance, and presently found what I have called dephlogisticated air, but without investigating the nature of it, and indeed, without being fully apprised of the degree of its purity. And though he says it *seems to be* more fit for respiration than common air, he does not say that he has

made any trial to determine how long an animal could live in it. He therefore inferred, as I have said that I myself had once done, that this substance had, during the process of calcination, imbibed atmospherical air, not in part, but in whole. But he extends his conclusion, and, as it appears to me, without any evidence, to all the metallic calces; saying that, very probably, they would all of them yield only common air, if, like *mercurius calcinatus*, they could be reduced without addition.

Lavoisier's attitude toward Priestley was, understandably, somewhat conflicted. While he once dismissed Priestley's work as "a fabric woven of experiments that is hardly interrupted by any reasoning," in a more generous moment he remarked, "I confess that I often have more confidence in Mr. Priestley's ideas than in my own." When, in 1777, he summed up the research leading to the definition of oxygen, Lavoisier acknowledged that "a part of the experiments contained in this memoir don't properly belong to me at all, perhaps, even, strictly speaking, there are some to which M. Priestley might claim the first idea; but, as the same facts have led us to diametrically opposed conclusions, I hope that, if anyone accuses me of having borrowed proof from the works of this celebrated physicist, at least no one will contest me on my entitlement to the conclusions." It is noticeable that Lavoisier's syntax grows rather less tortured as the sentence approaches surer ground: his strongest claim to priority was in interpretation.

But Priestley's criticism of the 1775 paper put a finger on a flaw in Lavoisier's interpretation as it then stood; in fact, metallic calces "imbibed atmospherical air" not in whole, but in part. Lavoisier spent another three years wrestling with this difficulty. He did studies which showed that respiration gradually replaced his "eminently breathable air" (oxygen) with fixed air (carbon dioxide). A more narrow examination of the calcination of mercury in the absence of charcoal revealed a third gas remaining after the oxygen had been absorbed into the calx: nitrogen, which Lavoisier named *mofette*. Lavoisier nailed this experiment to the balance sheet and was finally able to equate the weight gained in the calx with the weight lost from the air. The inverse operation also held; when Lavoisier heated the calx he had made to refine the mercury, the gas released (his eminently breathable air) was equal to the weight lost from the calx. Finally, by combining eminently breathable air with his *mofette*, ordinary atmospheric air was reproduced. The analysis and synthesis of air (which Scheele, wistfully, had mentioned) was thus achieved. "Here," Lavoisier declared, "is the most complete sort of proof at which one may arrive in chemistry: the decomposition of air and its recomposition."

Here, indeed, was a large advance. But the most radical achievement emerged from what at first appeared to be a byway. When frustrated by theoretical difficulties, Lavoisier was inclined to shift his attention to strictly delineated experiments whose purposes were as strictly limited. Now he turned to a close study of the role of air in the formation of acids. In April of 1777 he reported results of experiments with nitrous acid (which, not coincidentally, was important in the production of saltpeter for gunpowder). He dissolved mercury in nitrous acid, heated the resulting salt, and collected the gases and liquids released. When he reduced mercury nitrate by heat, he obtained water, *mofette* (nitrogen), and eminently breathable air. The weight of the refined mercury proved equal to the weight of the mercury he had started with.

Lavoisier was also able to demonstrate that his eminently breathable air was a component of carbonic acid, vitriolic acid, oxalic acid, and others. On September 5, 1777, he presented a paper to the Academy which contended that the principle that combined with metals to form calces was also a universal acidifying principle. In this lecture, he devoted a paragraph to clarifying his terminology: "I shall henceforward designate dephlogisticated air or eminently respirable air in the state of combination and fixity, by the name of acidifying principle, or if one likes better the same meaning in a Greek word, by that of *le principe oxygine.* That denomination will … put more rigor into my mode of expression, and avoid the ambiguities into which one would constantly risk falling if I used the word air."

In the final analysis, what Lavoisier had discovered was a *word.*

Oxy derived from the Greek word for acid; *gen* from the Greek for "beget." The use of the word *principle* suggests a strain of influence lingering from Paracelsus, Boyle, Becher, and especially Stahl, whose phlogiston was defined as an active principle rather than a material component of substances. In Lavoisier's emerging picture of things, however, oxygen *did* enter into chemical composition, as Rouelle's adaptation of phlogiston had done. For the moment, Lavoisier understood his eminently breathable air to be a combination of oxygen and matter of fire, though the fact that this notion did not provide a sufficiently clear distinction from Priestley's dephlogisticated air worried him. Eller's idea that air might be a combination of water with matter of fire also lingered in Lavoisier's thinking.

There were other small flaws in the first oxygen-based theory (oxygen is a component of some acids but not all, and so is not quite the universal acidifying principle Lavoisier posited it to be). But *le principe oxygine* was solid and sound enough to become a sort of axiom from which the rest of what-was-soon-to-be-called the "New French Chemistry" would evolve. The theory of elements toward which Lavoisier had been groping in his comparatively inchoate "system of elements" manuscript of 1772 now had, in oxygen, one firm leg on which to stand.

In the late 1770s, Lavoisier's laboratory in the Arsenal turned into a school—not only a workshop for the education of junior chemists but also an evolving school of thought. Jean-Baptiste Bucquet, who like Lavoisier had studied chemistry with both Rouelle and La Planche, came to teach beginning students at the Arsenal—Madame Lavoisier herself was a member of this class. When Bucquet died, Antoine-François de Fourcroy took over the role. The sense of cross-Channel competition was still very strong, and Lavoisier's laboratory, now clearly in the vanguard of French chemistry, exerted an international magnetism. Despite the fact that Lavoisier was extremely busy with the entirely separate careers in taxation and finance that he pursued at the same time, he was in the laboratory for some hours every day, and one whole day a week was devoted to experiments—with audiences generally welcome. Since the couple's private apartment was then in an Arsenal building as well, it was convenient for Madame Lavoisier to run a social salon to complement the program of experiments and demonstration, and her drawing room was frequented by the most illustrious scientists of Europe; Benjamin Franklin, during his tenure as United States ambassador to France was perhaps the most celebrated *habitué*.

Since 1773 Lavoisier had collaborated, off and on, with a junior member of the Academy of Sciences named Pierre-Simon de Laplace. Today Laplace is chiefly remembered as a pure mathematician, but he had a knack for designing scientific instruments that Lavoisier found valuable. In 1782, Lavoisier enlisted the help of Laplace in an experimental program meant to test his hypothesis that respiration is a mode of combustion that produces heat, or more precisely, that breaks down eminently breathable air to release its matter of fire.

By this time, however, Lavoisier had begun to depart from phrases like "matter of fire" and "igneous fluid." That sort of language had been used by Marat in the spurious interpretations of the experiments Lavoisier had helped to discredit in 1779 (though Marat's theories were too fantastic to constitute serious competition, Lavoisier took care to eliminate them nonetheless). In search of a term that had no flavor of the old alchemical lexicon, Lavoisier began to substitute the word *caloric* for "matter of fire." Caloric at least *sounded* like a concrete element—like oxygen—and at the time, Lavoisier believed that it was.

Whence the name of a new instrument Lavoisier and Laplace designed together: the "ice calorimeter." By this time Lavoisier was well versed in Joseph Black's theory of latent heat, which was an inspiration for the new machine. The innermost vessel of the ice calorimeter was contained in two nesting outer vessels, both packed with ice. The middle ring was insulated from the outside environment by the ice in the outermost vessel. Heat produced by the reaction in the inner-most vessel (be it combustion, mixture, or respiration) was measured in terms of the volume of melted ice drained from the middle ring.

Ice calorimeter experiments in the winter of 1782–83 sup-ported the idea that respiration was a form of combustion. Laplace and Lavoisier were able to show that both burning charcoal and a breathing guinea pig consumed oxygen and produced fixed air (carbon dioxide) while radiating heat that was measured by the volume of melted ice. The "Memoir on Heat" resulting from this experimental program was pre-sented to the Academy in June of 1783. The work was signifi-cant in the long term as a step in the direction of defining the mechanisms whereby mammals maintain a constant body temperature, and more generally in establishing techniques for the measurement of heat that would prove important to both chemistry and physics later on. What most acutely inter-ested Lavoisier, though, was a means to measure caloric, or fixed fire.

The net effect of many different experiments on various substances in numerous situations was leading Lavoisier toward the conclusion that calcination, respiration, and com-bustion were all variations of the same oxygen-consuming chemical process. Oxygen was the irreducible element discov-ered in the decomposition of air. Lavoisier looked to discover something analogous in the decomposition of another ele-ment among the Aristotelian four. If a component of the air could be fixed in bodies, why not a component of fire as well? "I will, of course, be asked what I mean by the matter of fire. I reply, with Franklin, Boerhaave, and one group of philoso-phers of antiquity in saying that the matter of fire or light is a very subtle and very elastic fluid that surrounds all parts of the planet we live on, which penetrates with greater or lesser ease all the bodies of which it is composed, and which tends, when free, to distribute itself uniformly in everything."

This hypothesis was an alternative to phlogiston theory in that it worked "without supposing that there is any material fire or phlogiston in combustible substances." Stahl's system held that burning objects lost phlogiston (a material with weight), while refined metals gained it—and yet calcined metals lost phlogiston while still somehow gaining weight. For years, Lavoisier had been homing in on that contradiction. Now he felt confident enough to say that if his own explanation worked out, "then the Stahlian system will find itself shaken to its foundations." What Lavoisier disliked most about Stahl's doctrine was that its proofs "inevitably fall into a vicious circle. They must reply that combustible bodies contain material fire because they burn and that they burn because they contain material fire. Obviously in the final analysis this is to explain combustion by combustion." Circular reasoning was the first thing Lavoisier had found objectionable in his earliest studies of chemistry.

The distinction between Stahl's entrenched doctrine and Lavoisier's emerging one remained somewhat difficult to grasp. What was the difference between phlogiston and Lavoisier's caloric or matter of fire? At this stage the difference had to do with location; whereas phlogiston was supposed to be found in solids, matter of fire was found in the air, and it combined with substances to increase their elasticity (an actual property of heat), a reaction that shifted solids to liquids and liquids to the state of vapor. Conceiving his matter of fire as a fluid, Lavoisier supposed that it dissolved various substances in the same way that water dissolves salts and acids dissolve metals. The new hypothesis did away with the difficult proposition that burning or calcining objects gained weight as they ostensibly lost phlogiston by positing instead that combustion and calcination involved fixation of the "base of pure air." Lavoisier was aware, however, that the proofs for this aspect of his evolving theory were not as yet overwhelming. The experiments he did with Laplace and the ice calorimeter were meant to establish a technique for isolating and quantifying caloric, but since heat (the real object of these experiments) is less substantial than Lavoisier believed it to be, to a certain extent it eluded them.

In the early summer of 1783, Lavoisier and Laplace succeeded with a much more conclusive experiment: the synthesis of water (H_2O). Hydrogen had been discovered by another English scientist, Henry Cavendish, in 1766—like oxygen it was a gas that burned, and Cavendish named it "inflammable air." Lavoisier had done experiments with burning hydrogen as early as 1774, believing that the reaction would fix air, but since he had no idea that water would be the product of this reaction, and because his apparatus used water to trap the gases, he could make no sense of his results.

Later, once he had conceived of *le principe oxygine* as the acidifying principle, Lavoisier expected that the combustion of hydrogen ought to form an acid. Not so. In 1781 Priestley, with another Englishman, John Waltire, noticed that burning hydrogen combined with atmospheric air left moisture on the inside of the flask—for Priestley, however, this result was proof that "inflammable air [hydrogen] is none other than phlogiston." In 1783 Cavendish synthesized water in a similar experiment and used an unusually convoluted phlogiston-based explanation to account for the result.

In June of 1783 Lavoisier and Laplace directed jets of hydrogen and oxygen into a pneumatic trough that used mercury instead of water to seal the bell jar, to better contain the water-soluble gases and to permit measurement of the liquid they believed the experiment would produce. The substitution of mercury for water in the trough corrected the flaw in Lavoisier's previous effort to measure the combination of hydrogen and oxygen; by now he probably anticipated that water might be the result. He and Laplace burned the mixture of hydrogen and oxygen—establishing the correct proportion by looking for the brightest flame—and collected water from the glass receptacle sealed by the mercury. The subsequent measurements were not perfectly precise, but Lavoisier was confident enough to declare on June 25 that "if one burns a little less than two parts of aqueous inflammable air under a bell jar with one part of vital air, assuming that both are perfectly pure, the totality of the two airs is absorbed and one finds on the surface of the mercury a quantity of water equal in weight to that of the two airs used."

In those years the Montgolfier brothers were experimenting with hot-air balloons, while the physicist Jacques Charles competed using hydrogen-filled balloons. Lavoisier took an interest in these ballooning ventures, and along the way looked into various methods for mass-producing hydrogen. Jean-Baptiste Meusnier, a physicist and engineer whom Lavoisier had consulted on various technical problems since his early studies of the Parisian water supply, helped him with the latter effort.

In 1785 Lavoisier returned to experiments on the composition of water. With the help of Meusnier he designed extremely impressive new instruments: two gasometers, each connected to a combustion flask. With their volumes precisely regulated by the gasometers, hydrogen and oxygen flowed into the combustion vessel, where they were ignited by an electric spark and synthesized into water. The procedure for the analysis was only slightly less spectacular: water dripped into a red-hot iron gun barrel decomposed into oxygen and hydrogen; the latter was collected in a water pneumatic trough. Since Meusnier and Lavoisier had evolved this procedure to accumulate hydrogen for balloons, they let the oxygen dissipate.

The price of the apparatus for these experiments was equally spectacular. Just one of the two gasometers used cost the equivalent of 250,000 U.S. dollars in the twentieth century. Arguably, they were worth the money, for they were minutely accurate, marvelously efficient, and beautifully designed and realized—the best the emerging Industrial Revolution had to offer. By extreme contrast, the over-awing aspect of Lavoisier's equipment recalls the more retiring and less prosperous Scheele, who lacked an adequate burning glass to consolidate his discovery of oxygen. Demonstrations on the scale of Lavoisier's analysis and synthesis of water were enough to cow any possible competition completely out of the field.

Lavoisier had seen the value of dramatic presentation in Rouelle's sometimes explosive public demonstrations, and the lesson was reinforced when he worked outdoors in the center of Paris with the huge Tschirnhausen burning lenses. Observed by an international audience of more than thirty scientists, his analysis and synthesis of water was as majestic in

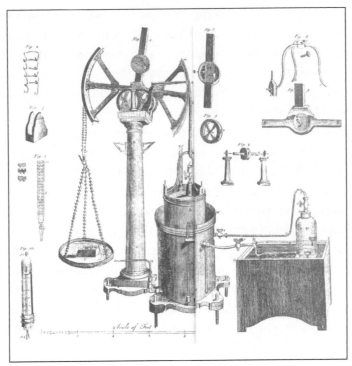

Lavoisier's gasometer, drawn by Madame Lavoisier

its way as the vision of hydrogen balloons arising from the
Parisian Champ de Mars. It had the intended effect on an
English observer, Arthur Young, who reported, "It is a noble
machine. M. Lavoisier, when the structure of it was com-
mended, said, *Mais oui, Monsieur, et même par un artiste
Français!* with an accent of voice that admitted their general
inferiority to ours."

If Lavoisier really meant to be self-deprecating, he could
well afford the gesture, since the analysis and synthesis of
water proclaimed, on an imperial scale, his nation's absolute
mastery of what was now known as the "New French Chem-

istry." As for the scientific conclusion, it could be succinctly stated: "water is not an element, it is on the contrary composed of two very distinct principles, the base of vital air and that of hydrogen gas; and . . . these two principles enter into an approximate relationship of 85 to 15 respectively."

With his work on minerals and metallic ores, with the decomposition and recomposition of air, and with the analysis and synthesis of water, Lavoisier had—literally—taken apart three of the four Aristotelian elements. These discoveries, and his increasingly confident placement of them in an expanding theoretical context, were enough to make good on the promise of a revolution in chemistry he had made to himself ten years before.

Among the elements of antiquity, only fire still resisted his analysis. Understandably, from the modern point of view, he was unable to get a material grasp on it. However, by 1785 Lavoisier was fully prepared to dismantle phlogiston. In fact, most of the experimental work toward that end had already been done. Henceforward, the overthrow of phlogiston theory would be a political effort.

A print of David's portrait of the Lavoisiers

IV

The Chemical Revolution

In 1788, David, the prince of eighteenth-century neoclassical painting, delivered a full-length portrait of Monsieur and Madame Lavoisier to his clients. At first glance, it is Marie-Anne Lavoisier who dominates the scene. Monsieur Antoine Lavoisier is, in fact, just slightly in her shadow, his head raised and turned away from his paperwork, as if his attention has just now been distracted by his wife's arrival at his side, her left hand lightly resting on his shoulder and her right hand grazing the lace around the wrist of Lavoisier's writing hand, knuckles bent against the rich red cloth on the table. Lavoisier is richly but soberly dressed, all in black save the obligatory white lace at his wrists and throat and the silver buckle of his shoe. Indeed, he appears to be "making a leg," in the eighteenth-century courtier's fashion, but from the waist up he conveys an intellectual concentration barely broken, in which he will soon again be submerged.

His eyes are only on his wife; it is she, alone, who turns her gaze outward toward the viewer—as if she had just noticed us, but is in no way disconcerted by our presence. Her dress, floor length in simple white muslin, bound with a blue sash, captures and anchors the light that pours in from the upper left of the frame. David's composition comes to rest on her face, which, full of calm, meets our regard. Marie-Anne Lavoisier is in her early thirties; the first blush of her youth has faded, but it is still an attractive face, whose features express intelligence, strength of purpose, and the patience to persistently apply those first two qualities.

David was, among other things, an active painter of the theater, and the Lavoisier portrait, though far from operatic, reads as a theatrical scene. The Lavoisiers have been captured in motion, though their motion is quiet and restrained: the image is a quick-frozen frame of a marriage that has subordinated its intimacy and affection to the needs of an intellectual partnership. Marie-Anne rests gently on her husband's shoulder, in a familiar domestic pose, but her right arm is more forcefully extended to the table, reinforcing Lavoisier's hand, which holds the pen. His left hand rises in an interrogative gesture, his lips are slightly parted, as if to speak. He is looking to his wife for support, advice, the answer to a question— at any rate, for something. While he ignores us, she gives us a steady but dismissive gaze. We sense that in the next moment the two of them will have returned to their work.

David has caught the Lavoisiers in their prime—at the height of their accomplishments and of the recognition those accomplishments had won. Antoine Lavoisier had reached his greatest prominence in public service; he was a force in Parisian city planning, in the French national economy, in agricultural reform, and in an emerging chemical industry. In pure science he was now generally recognized as the theoretician who had organized what everyone now termed a chemical revolution. Marie-Anne was his steady supporter in all his scientific advances. Beginning in 1777, she and a few other select students took courses in chemistry from Jean-Baptiste Bucquet in Lavoisier's Arsenal laboratory, thus rendering herself competent to serve her husband as a lab assistant, as an editor of his own memoirs and reports, and as a translator of chemical treatises from other languages into French. The David portrait seems to show her assisting him with the correction of some such manuscript.

In 1786 Marie-Anne had begun courses in drawing and painting with David himself. Two drawings survive from this tutelage, both bearing David's handwritten comments. On a painstaking charcoal study of a bust of Antinoüs, he inscribed the phrase "very good" three times. A copperpoint study of a classical nude is marked with the sentence, "Up to now, I could not be happier." David's praise was not wholly disinterested; his price for the Lavoisier portrait was seven thousand livres—a higher rate than that obtained by painters working for the king of France. Madame Lavoisier's surviving artworks reveal her to be competent but not remarkably gifted. The one painting she is known to have completed—a portrait of Benjamin Franklin presented to him in 1788—drew this response from its subject: "Those who have seen it declare that the painting has great merit and that it is worthy of consideration, but what makes it dear to me above all is the hand that held the brush."

In his portrait of the Lavoisier couple, David places an artist's portfolio on an armchair at Marie-Anne's back, indicating that her competence as a draftsman was an important part of her repertoire. Antoine Lavoisier often expressed his evolving ideas in crude sketches of his own; Marie-Anne's refinements of these extremely rough drawings may well have helped to clarify his thoughts. Lavoisier sketched equipment he wanted to be built; his wife supplied much more meticulous and accurate images of this apparatus once it had been created, served its purpose, and proved its point. The thirteen plates in the 1789 edition of the *Traité élémentaire de chimie* are hers.

Two sepia drawings, circa 1790, provide a unique graphic image of the scene in the Arsenal laboratory at the peak of its productivity (and also the only record of the instruments used in experiments on human respiration that Lavoisier carried out at this time). David's influence is evident in these compositions, which are based on neoclassical symmetry. Madame Lavoisier portrays her husband in quasi-heroic poses, as leader of his several assistants and master of his human subject (Armand Seguin, a younger chemist whose head is sealed with a mask and hood that route his breath to the apparatus via a long tube). She includes herself in both drawings, in her quite consequential role of recording the experimental data. In both she is peripheral to the composition, but at the same time essential to its balance.

In these respects, Marie-Anne Lavoisier seems to have been a near-perfect wife—though never a mother—an ideal helpmeet and research partner. The American Gouverneur Morris, who spent some time with the Lavoisiers in the fall of 1789, recorded, "Madame appears to be an agreeable woman. She is tolerably handsome, but from her manner it would seem that she thinks her forte is the understanding rather than the person." If this praise seems a little pallid, Morris pursued the acquaintance with a certain gallantry nonetheless. "When she told me she had no children, I jokingly chided her for her idleness, but she replied only that she had been unlucky."

The one asymmetry in the carefully controlled image that Marie-Anne Lavoisier presented to the world is that in 1781 she began a long affair with Lavoisier's friend and colleague, Pierre-Samuel Dupont, who was, like Franklin and many other scientific luminaries, a frequent visitor to the Lavoisier salon. Fourteen years older than his wife, Lavoisier seems to have become in some respects an absentee husband. Marie-Anne preferred Paris to the country estates, where her husband spent increasing amounts of time; Dupont was more worldly, more charming to women, and probably more carnally inclined than the abstracted chemical theoretician. Such affairs were common in the day, and eighteenth-century French society tolerated them. But Marie-Anne Lavoisier seems to have been unusually discreet about hers. In the 1780s she attended scientific lectures offered in the evenings at the Tuileries, and had the curious habit of arriving by carriage but departing on foot—and the way back to their apartment in the Arsenal was long. An observer, baron de Frénilly, attributes this behavior to penuriousness, but given the Lavoisiers' very large wealth, that explanation seems unlikely; biographers now think it more probable that her solitary nocturnal promenades were a means of disappearing from the eyes of her friends into the arms of her lover. The relationship remained secret during her lifetime, and was discovered afterward from letters written by Dupont.

If we look at David's portrait with this knowledge, do we find suspicion in Lavoisier's raised eyes? The affair had been going on for seven years, but Marie-Anne's discretion may

Drawing by Madame Lavoisier of one of Lavoisier's experiments on respiration, this time with a resting subject.

have protected her husband from any inkling of its existence. His expression and attitude in the portrait are more often interpreted as a variation on the conventional theme of the poet inspired by his muse. Whether Lavoisier enjoyed mistresses of his own during these highly successful years has not been recorded. It would have been quite conventional if he had, but there is no evidence that he did; he seems to have had little or no inclination for adventures of that kind. And he was so extremely busy that he hardly had time for dalliance.

THE OTHER ACCOUTREMENTS of the scene in the David portrait are all scientific; perhaps at the behest of his subjects, the artist took care to represent the instruments used to accomplish the chemical revolution, many of which Lavoisier had a hand in designing. On the floor, near Lavoisier's elegant shoe, is a flask that closely resembles the one actually used in the synthesis of water. Partially concealed by the glass of this vessel is the brass hydrometer that Lavoisier had remodeled for the measurement of the density of fluids. Biographer Jean-Pierre Poirier notes that these placements must have been imposed by the artist, since the meticulous Lavoisier would never have left his instruments so carelessly strewn on the floor.

The gasometer pictured next to Lavoisier's inkwell is an earlier model than those used in the definitive synthesis of water in 1785, and also a lot smaller—so more convenient for inclusion in this parlor setting. Its presence, along with the narrower tube of mercury beside it, is meant to recall to the observer the discovery of oxygen's release from *mercurius calcinatus*. At the edge of the image, partially eclipsed by the frame, is a Lavoisierian version of the pneumatic Hales device: a glass bell of water inverted in a porcelain bowl. Of the instruments most vital to Lavoisier's crucial experiments, only the balance scale is missing; perhaps David did not find it visually appealing, or perhaps it was in service elsewhere.

The text that Lavoisier has momentarily been distracted from correcting is supposed to be his *Traité élémentaire de chimie*, and the portfolio at Madame Lavoisier's back is thought to contain the thirteen engraved illustrations of that work, which she had completed under the tutelage of David. The instruments posed on the table and the floor are arranged in the same order that the key experiments in which they were used are presented in *Traité*. The David portrait was intended for exposition in the Salon of 1789, the same year that *Traité élémentaire de chimie* was published; and the painting is as authoritative a visual document as the treatise is a verbal one. The text and the image reinforce each other—sending the message in all available media that the chemical revolution has been consolidated, with Antoine Lavoisier at its head.

THE BASTION THAT the chemical revolution had first needed to overthrow was phlogiston. Lavoisier had lightly probed its fortifications many times prior to 1785. In the early 1780s, following his success with the composition of water, his attacks became more aggressive: "This entity, introduced into chemistry by Stahl, far from having brought light to bear upon it, seems to have created an obscure and unintelligible science for those who have not made a highly specialized study of it; it is the *deus ex machina* of the metaphysicians: an entity which explains everything and which explains nothing." At this stage of chemistry's development, the critique was quite reasonable: the theorists building on Stahl's hypothesis had invested phlogiston with so many contradictory qualities that Lavoisier could easily feel justified in denouncing it as a *deus ex machina*. To say further that phlogiston was the *deus ex machina of the metaphysicians* was to accuse it of being not a rational but a magical solution. Phlogiston had turned into a facilitating mechanism for the sort of circular reasoning that had aroused Lavoisier's suspicions from his very first explorations of chemistry in his youth.

Here the lines of battle were drawn. In June of 1795, Lavoisier read to an assembly of the Academy of Sciences from his *Réflexions sur la phlogistique* (Reflections on Phlogiston). "Chemists have made of phlogiston a vague principle which is not at all rigorously defined," he accused, "and which consequentially adapts itself to any explanations into which they wish to make it enter; sometimes this principle is heavy, and sometimes it is not; sometimes it is free fire, and sometimes it is fire combined with the earthy element; sometimes it passes through the pores of vessels, sometimes these are impenetrable to it; it explains all at once causticity and non-causticity, translucency and opaqueness, colors and their absence. It is a veritable Proteus, which changes its form every second. The time has come to bring chemistry back to a more rigorous manner of reasoning . . . to distinguish what is fact and observation from what is systematic and hypothetical."

In the last sentence one glimpses again a recurring paradox in Lavoisier's thinking; though his greatest achievements were in theory, he was always fundamentally mistrustful of a system's tendency to dismiss or distort experimental facts for the sake of its own internal consistency. Indeed, in their early stages Lavoisier's own theories were sometimes guilty of this fault. In his 1777 memoir on combustion, he had struggled openly with the problem. "Dangerous though the spirit of systems is in physical science, it is equally to be feared lest piling up without any order too great a store of experiments may obscure instead of illuminating the science: lest one thereby make access difficult to those who present themselves at the threshold: lest, in a word, there be obtained as the reward of long and painful efforts nothing but disorder and confusion. Facts, observations, experiments, are the materials of a great edifice. But in assembling them we must not encumber our science. We must, on the contrary, devote ourselves to classifying them, to distinguishing which belong to which order, to each part of the whole to which they pertain."

By the summer of 1785, in his *Réflexions sur la phlogistique*, Lavoisier felt confident enough "to show that Stahl's phlogiston is imaginary and its existence in the metals, sulphur, phosphorus, and all combustible bodies, a baseless supposition, and that all the facts of combustion and calcination are explained in a much simpler and much easier way without phlogiston than with it." That is to say that as he built his "facts, observations, experiments" into the edifice of a new order, a superior (because simpler) explanation of the phenomena under discussion would emerge.

With statements like these, Lavoisier appeared to be cutting with Occam's razor, though whether he had direct contact with that "principle of parsimony" is uncertain. Occam's razor was not generally part of the toolkit of eighteenth-century scientists, but Lavoisier's refutation of phlogiston theory is a classic example of its application. In the fourteenth century, William of Occam had proposed that all reasoning should proceed from the minimum number of assumptions necessary—a concept that curbed theory's innate tendency to embroider itself. One should remember that Stahl too had used the idea embodied in Occam's razor when he simplified myriad alchemical fantasies about combustion with his assumption of phlogiston. And Lavoisier always made it clear that his attack on phlogiston was not meant as an attack on Stahl. Lavoisier was cutting more dross from chemical theory by showing that chemistry could be clearer and simpler without phlogiston than with it.

In the conclusion of *Réflexions sur la phlogistique*, Lavoisier remarked (with the *sangfroid* that characterized many of his highest-risk maneuvers), "I see with much satisfaction that young men, who are beginning to study the science without prejudice, and geometers and physicists, who bring fresh minds to bear on chemical facts, no longer believe in phlogiston in the sense that Stahl gave it and consider the whole of this doctrine as a scaffolding that is more of a hindrance than a help for extending the fabric of chemical science." To make this statement before the Academy of Sciences was something like throwing his hat over the walls of an enemy fortification—so committing himself to retrieve it with glory or perish in the attempt. In fact, in the summer of 1785 Lavoisier had not yet made any influential converts to what would soon become notorious as the "antiphlogistic" doctrine.

A point of vulnerability for Lavoisier's antiphlogistic theory was that critics could dismiss it (along the lines of Marat's diatribe) as no more than a variation on the phlogiston theme. Lavoisier's theory did retain the concept of caloric, or matter of fire, alongside the material fact of oxygen—and here the main difference between his theory and Stahl's was that the latter located matter of fire in combustible materials, whereas Lavoisier installed it in oxygen. In fact, many critics were quick to claim that Lavoisier's *Réflexions sur la phlogistique* had done little more than recite Stahlian theory backward, like holy scriptures inverted in a Black Mass.

Organized opposition came from England, where Priestley, not without justification, was still grumbling over Lavoisier's preemption of the discovery of oxygen. In 1787 the Irish chemist Richard Kirwan, a colleague of Priestley's in the Royal Society, published *An Essay on Phlogiston and the Constitution of Acids*. As Lavoisier had tended to do in Priestley's case, Kirwan bowed respectfully in the direction of Lavoisier's experiments, then launched an attack on his theory. Kirwan's counterinterpretation of Lavoisier's results (several of which, as the Irishman tartly pointed out, had been previously obtained by Priestley, Cavendish, and Adair Crawford) was that the fact that metals in contact with acids exhaled combustible gases was proof of the existence of phlogiston. Kirwan decided to identify hydrogen with phlogiston, and spent much of his volume trying to establish the presence of hydrogen/phlogiston in all combustible substances. He finished by acknowledging that Lavoisier's "antiphlogistic hypothesis" was "recommendable in its simplicity," while still discrediting it as "more arbitrary in its application, and less countenanced by the general rules of philosophic reasoning" than Stahl's phlogiston theory.

Kirwan's counterattack was successful, at first, in England, Germany, and Sweden. The first response from the Lavoisier camp was for Marie-Anne Lavoisier to translate the work into French. This move, though counterintuitive at first glance, proved to have been astute in the endgame.

Laplace, the mathematician who had spent several years helping Lavoisier design equipment and calculate results, was an early convert to the new, antiphlogistic doctrine. Other mathematicians came with him: Jacques Cousin and Alexandre Vandermonde. Antoine Fourcroy, a young chemist who often worked in Lavoisier's Arsenal laboratory and participated in the theoretical discussions there, was convinced early, or half-convinced.

On the death of Bucquet (Madame Lavoisier's chemistry tutor), Fourcroy took over Bucquet's chemistry courses. For a time he tried to teach Lavoisier's theory in tandem with Stahl's, but gradually he allowed the new theory to replace the old one. As Rouelle had once done, Fourcroy offered chemistry courses to the general public in the Jardin du Roy and proved more influential than anyone else in popularizing the antiphlogistic doctrine through his teaching.

Claude-Louis Berthollet, an influential chemist who had

worked closely with Lavoisier in both the ballooning projects and the gunpowder trade, made a public declaration of support for the new theory as well, and his influence helped bring others to Lavoisier's side, including Gaspard Monge and Jean-Antoine Chaptal. Guyton de Morveau, then at work on a dictionary of chemistry, wavered between the two theories. Lavoisier, Monge, and Fourcroy traveled to Dijon to try to win him to their side, while Kirwan, by letter, struggled to keep him loyal to phlogiston.

Lavoisier's opposition remained powerful. Older chemists like Macquer were reluctant to relinquish phlogiston. Jean-Claude de Lamétherie, an ardent opponent of the new chemistry, had in 1785 become editor of the most influential scientific publication in France, Jean-François Pilatre de Rozier's *Journal de physique*. Richard Kirwan's retort was rallying phlogiston supporters all over Europe and especially in Germany and England, where opposition to the French innovations had a nationalistic dimension.

Early in 1788 Lavoisier and his allies responded with a maneuver of Machiavellian sophistication. They themselves published a French edition of Kirwan's *Essay on Phlogiston*—accompanied by a multifaceted refutation. To this volume, Lavoisier, Berthollet, Fourcroy, Monge, and Guyton de Morveau contributed articles disproving the existence of phlogiston in their various areas of expertise. For another year, the echoes continued to rebound: in 1789 Kirwan published a new edition of his book that included the French commentaries and an effort to rebut them, which, though not without some effective sarcasm, was clearly losing energy. In 1791 Kirwan finally gave up on phlogiston; most other chemists had already done so.

———

LAVOISIER'S CAMP HAD opened another offensive in 1787; the new front was linguistic, and the victories here, though slow to be realized, were the most far-reaching of all. The project was to create a new terminology to express the New French Chemistry, and the process of doing so forged and tempered the core of the antiphlogiston alliance: Lavoisier, Fourcroy, Berthollet, and (finally) Guyton de Morveau. By extension of the same reasoning that had birthed the name of oxygen (at once the project's first neologism and first inspiration), these four set out to reform the whole chemical nomenclature.

Clear expression had always been as important to Lavoisier as lucid reasoning, procedure, and analysis. For him, all of these operations were intertwined. "Languages don't have the sole object (as is commonly believed) of expressing ideas and images through signs; they are, additionally, veritable analytic methods, by whose aid we proceed from the known to the unknown. . . . An analytic method is a language; a language is an analytic method, and these two expressions are, in a certain sense, synonyms."

This statement is an explicit and undoubtedly conscious echo of another of the eighteenth-century Lumières, Étienne Bonnet de Condillac, who had written, "The art of speaking, the art of writing, the art of reasoning and the art of thinking are, at bottom, only one and the same art. In effect, when one knows how to think, one knows how to reason, and to speak well and to write well nothing remains but to speak as one thinks and to write as one speaks." This synergistic picture of the intellectual faculties had a tremendous influence on teachers, philosophers, and scientists at the time when Lavoisier and his three comrades commenced the invention of the new chemical nomenclature. In April of 1787, Lavoisier introduced the idea to the Academy: "The word ought to bring about the birth of the idea; the idea should depict the fact; they are three imprints from the same stamp. And since it is words that preserve and transmit ideas, the result is that it would be impossible to improve science without improving its language."

In his introduction to *Méthode de nomenclature chimique*, Lavoisier explained how, according to Condillac, "the language of algebra could be translated into the vulgar tongue and reciprocally; how the progress of the understanding and judgement is the same in both; and how the art of reasoning is the art of analysing." Here was a pathway to the mathematization of chemistry that Lavoisier had been seeking for most of his career. "To analyse, then," Condillac had written in his *La logique*, first published in Paris in 1780, "is nothing other than to observe the qualities of an object in a successive order so as to give them, in our mind, the simultaneous order in which they exist outside of it. Therefore analysis, which people think is known only to the philosophers, is actually known to everyone; and I have taught the reader nothing. I have simply made him note what he continually does." In Condillac's vision, the method of analysis was universally applicable, to morality and politics as well as to anything and everything in the sphere of the sciences and natural philosophy, though the *Méthode de nomenclature chimique* did not try to extend it so far.

What Lavoisier recognized, however, was that the method of analysis "which is necessary to be introduced into the study and teaching of the science of chemistry, is closely connected with the reformation of the nomenclature." And he saw that the right expression of the analytic method would have a compelling force: "A well-made language, a language in which we have seized the natural and successive order of ideas, brings with it a swift and necessary revolution in the way of teaching; it does not permit those who profess chemistry to stray from the march of nature; they must either reject the nomenclature or irresistibly follow the route it has marked."

Irresistible indeed—any user of the new French nomenclature would have no choice but to accept the antiphlogiston theory on which it was based. "We are very far," Lavoisier's introduction admitted, "from knowing the entire extent of all parts of chemistry, but provided that it has been undertaken upon good principles, provided that it is rather a *method of naming* than a *nomenclature* it must naturally adapt itself to future discoveries, and indicate beforehand, the place and name of such new substances as may be found out, and we can never require more than some local and particular reformations." The new chemical lexicon would not be an arbitrary assignment of names to substances, as most alchemical terminology had been, but a deployment of language to express chemical relationships—language as an analytic tool.

THE SIMPLE, UNDECOMPOSABLE substances were the first to be defined. At the top of this list was the elusive matter of fire now redefined as "caloric"—the last vestigial trace of phlogiston theory in the new chemical system, for Lavoisier and his partners had yet to recognize that heat was not matter but energy (though by 1789 he would begin to say "we are not obliged to regard [caloric] as a real substance"). Tools of analysis had revealed the other irreducible and thus elemental substances: oxygen, hydrogen, azote (nitrogen), sulfur, phosphorus, iron, gold, and so on. Compound substances were expressed by neologisms in which the suffix described a quality of their composition—thus, the suffix of *calcium nitrate* reveals a higher oxygen content in this salt than in its cousin compound *calcium nitrite*.

In his youth, Lavoisier had been exposed to Linnaeus's system of biological classification, when he botanized with Bernard de Jussieu. Linneaus classified plants by a binomial technique in which one word indicated genus and the other species. Guyton de Morveau had begun an effort to import this system into chemistry as early as 1775. To the comparatively static Linnaean system of classification, Lavoisier now added Condillac's sense of naming as the application of an analytic method.

Where previous names and terms in chemistry had been superimposed on observable facts by theory, the terms of the new nomenclature were derived from the analysis of substances and for the most part founded on explicit, well-confirmed laboratory results. Furthermore, the new definitions of simple, elemental substances brought centuries of investigation of the composition of matter to a near-definitive conclusion. Henceforward, the central subject of the science would be not composition but change.

With the aid of the new nomenclature, combined with the principle of conservation of matter, chemical analyses could be worked out in algebraic terms. "I suppose that I have to analyze a salt, of which I know neither the acid nor the base," Lavoisier wrote in a 1787 memoir.

> I introduce a known weight of this salt into a retort; I pour vitriolic acid over it and distill; I obtain nitrous

acid in the receiver; I conclude that the salt is niter.

But what is the mechanism of reasoning which has led me to this result? An instant of reflection will soon make it known. It is first of all clear that if I wished to make an exact calculation of quantities I had to suppose that the weight of the materials employed was the same before and after the operation and that only a change of modification took place. I therefore mentally set up an equation in which the materials existing before the operation formed the first member, and those obtained after the operation formed the second; and it was actually by the resolution of this equation that I successfully obtained the result.

By dint of this algebraic procedure, Lavoisier actually identified two unknown variables, nitric acid and fixed alkali, on his way to identifying the mystery salt as niter—an introduction of mathematical precision into chemistry that would have gladdened the heart of his old chemistry teacher Lacaille.

DESPITE THE ASSIDUOUS courtship of Lavoisier and his allies, Guyton de Morveau remained a holdout for phlogiston for quite some time. Apparently it was the nomenclature reform that finally tempted him over to the new chemistry it expressed. Guyton had attempted something similar in 1782—an early effort to create classifications on the Linnaean model for chemistry. He already understood the goal, and becoming the fourth contributor to *Méthode de nomenclature chimique* gave him a way to achieve it. Lavoisier had put forward the notion that chemists "must either reject the nomenclature or irresistibly follow the route it has marked." Guyton's conversion from phlogiston to oxygen theory was an early and important proof of that hypothesis; others would follow, though not without difficulty.

In June of 1787, Academy members Baumé, Cadet, Jean Darcet, and Balthazar Sage delivered a report on the new nomenclature that attacked Lavoisier's oxygen-based theory and defended phlogiston. Baumé, an unsuccessful player in the discovery of oxygen, was perhaps still irked by Lavoisier's having beaten him to that punch. A similar critique from Jean Claude de Lamétherie appeared in the *Journal de physique*. Like many who resisted the new theory and nomenclature, Lamétherie found the very idea of systematically creating a new terminology to be objectionable: "there is nothing else than the general consensus, which is called *use*, which has the authority to change some terms and replace them with others." The opposition of this publication was so problematic for Lavoisier and his supporters that they eventually founded a competing journal: *Annales de chimie*.

Opponents of the new terminology did not by any means miss the not-especially-well-hidden agenda of *Méthode*: anyone using the nomenclature would also be compelled, as a side effect of adopting this language, to accept in the bargain the theory it expressed. Moreover, the Frenchification of the new terminology was especially obnoxious to commentators outside of France. Thus, Joseph Black: "These names have evidently been contrived to suit the genius of the French language, in the first place, and then to have been transferred into the Latin words; or the words which were meant to be used in the Latin language have been coined from the French ones. When changes are thus made in the names of things which are familiar to us, I believe most people find them *disgusting* at first, on account of the shock and derangement which they give to the habits we had formed before. These latinized French words appeared to me at first very *harsh* and *disagreeable*." Hostile as it was, Black's commentary was also perceptive; the linguistic "genius" he noticed was shaping the neoclassical literature of Corneille and Racine as well as the texts of the Encyclopédistes; the exquisitely rational instrument that the French language was becoming was ideal for the architecture of theory. In fact, Black was reacting specifically against "that *itch for theory*," otherwise termed "a rage for system." Henry Cavendish also objected to "the impropriety of systematic names and the great mischief which will follow from [Lavoisier's] scheme if it should come into use." Across the Atlantic, Thomas Jefferson predicted failure for the new French chemical language: "it is premature, insufficient, and false . . . and upon the whole I think the new nomenclature will be rejected after doing more harm than good." Franklin, to whom Lavoisier had eagerly sent a copy of *Méthode*, chose to send his thanks to Madame Lavoisier instead of her husband, with the distinctly noncommittal remark, "It must be a very useful book."

It was peculiar (and no doubt rather frustrating to the authors) that the new nomenclature won over French mathematicians and physicists more easily than French chemists. In England, Lavoisier's first convert was a theologian, Edward King. Chaptal, who adopted the new nomenclature early, looked about himself and observed that most of his contemporaries "want the revolution to come about gradually." But Chaptal, like Fourcroy, was a teacher, and the convenience of the nomenclature for teaching would prove to be a very strong survival trait; even Joseph Black began teaching it, in the end.

In 1786, when Guyton de Morveau contributed the first of his chemistry volumes to the *Encyclopédie méthodique*, he was still a phlogiston loyalist. By 1789, his conversion to the antiphlogiston theory and terminology was so complete that in the introduction to the second volume he made a not-very-plausible claim that the whole nomenclature project had been his idea in the first place. This text drove a stake through the heart of phlogiston theory, which (Guyton now wrote) had been useful in its day—before it was definitively refuted by advances in pneumatic chemistry in the late 1770s and early 1780s. Guyton reprinted an updated version of the new chemical terminology (thus installing it in a standard reference work) and sealed it in place with a reiteration of Lavoisier's point that what had been created was "*a method of naming rather than a nomenclature.*" Lavoisier's prediction that this method would prove permanently valid for naming new substances might have seemed rash at the time that he made it—but, in fact, the same method *is* still used today.

THE NEW CHEMICAL nomenclature and the French translation (and refutation) of Kirwan's challenge to oxygen theory were the two great propaganda tools of the chemical revolution. In her preface to her translation of Kirwan's work, Marie-Anne Lavoisier declared with a distinct satisfaction, "The study of chymistry becomes daily more general, and its advances appear to be more particularly rapid since a philosopher, well known for the scrupulous attention with which he makes his experiments, and the philosophical spirit which has directed his observations, has formed a new theory, in which nothing is admitted but established facts." With these phrases, Madame Lavoisier presented a verbal portrait of her husband as flattering in its way as David's painting of the increasingly famous couple.

But the capstone of the theoretical arch that Lavoisier and his allies were constructing was Lavoisier's *Traité élémentaire de chimie*—the work whose manuscript was rendered among the other icons included in David's portrait. First published in 1789, this text consolidated the achievements of the chemical revolution just as the French political revolution began.

Lavoisier's first interest in chemistry had been awakened by his reaction against the confusing manner in which it was taught, when he encountered it as a student at the Collège Mazarin. His first impulse had been to reform the method of *teaching* chemistry. Along his way toward that end he had also radically altered the entire content of the subject. Nonetheless, his *Traité élémentaire de chimie* was designed and presented as a textbook.

His preface included a somewhat abstract critique of previous methods of teaching the sciences:

> It is by no means to be wondered, that, in the science of physics in general, men have often made suppositions, instead of forming conclusions. These suppositions, handed down from one age to another, acquire additional weight from the authorities by which they are supported, till at last they are received, even by men of genius, as fundamental truths.
>
> The only method of preventing such errors from taking place, and of correcting them when formed, is to restrain and simplify our reasoning as much as possible. . . . We must trust to nothing but facts: These are presented to us by Nature and cannot deceive. We ought, in

every instance, to submit our reasoning to the test of experiment, and never to search for truth but by the natural road of experiment and observation. Thus mathematicians obtain the solution of a problem by the mere arrangement of data, and by reducing their reasoning to such simple steps, to conclusions so very obvious, as never to lose sight of the evidence which guides them.

This passage embodies Lavoisier's fondness for mathematical precision, his determination to give priority to experimental evidence, and his highly evolved mistrust of received ideas. "Thoroughly convinced of these truths," he goes on, "I have imposed upon myself, as a law, never to advance but from what is known to what is unknown, never to form any conclusion which is not an immediate consequence necessarily flowing from observation and experiment; and always to arrange the facts, and the conclusions which are drawn from them, in such an order as to render it most easy for beginners in the study of chemistry thoroughly to understand them." Straightforward as it seems, this sentence is also a rather close paraphrase of a passage in Condillac's *Logique*, and thus carries with it all the implications of Condillac's universal application of analysis. Lavoisier certainly meant his book to be useful to beginning students of chemistry, but he was also suggesting that *all* chemists would have to begin all over again with a fresh conception of the whole subject—the one Lavoisier had begun to formulate when he finished his own education in chemistry twenty years earlier.

"It is not to the history of the science, or the human mind, that we are to attend in an elementary treatise: Our only aim ought to be ease and perspicuity, and with the utmost care to keep every thing out of view which might draw aside the attention of the student; it is a road which we should be continually rendering more smooth, and from which we should endeavor to remove every obstacle which can occasion delay." Among the obstacles so removed was any reference to competing views and theories of chemistry; Lavoisier was frank in the admission "that I have stated only my own opinion, without examining that of others." If his contention that "long dissertations on the history of the science, and the works of those who have studied it" would only have produced "a work the reading of which must have been extremely tiresome to beginners" was somewhat disingenuous, the streamlining effect of his choices of what to include and exclude was certainly significant.

The preface concludes with two substantial quotations from Condillac—each extremely potent for Lavoisier's reconception of chemistry. The first of these "observations" describes the path of error, and though, as Lavoisier notes, it was "made on a different subject," it serves as an abstract analysis of how erroneous principles got themselves ensconced in chemistry—from the alchemical period up until Lavoisier's own time: "Instead of applying observation to the things we wish to know, we have chosen rather to imagine them. Advancing from one ill founded supposition to another, we have at last bewildered ourselves amidst a multitude of errors. These errors becoming prejudices, are, of course, adopted as principles, and we thus bewilder ourselves more and more. The method, too, by which we conduct our reasoning is as absurd; we abuse words which we do not understand, and call this the art of reasoning."

To cure the habitual substitution of prejudice for principle,

Condillac proposed this solution: "when errors have been thus accumulated, there is but one remedy by which order can be restored to the faculty of thinking; this is, to forget all that we have learned, to trace back our ideas to their source, to follow the train in which they rise, and . . . to frame the human understanding anew." Here was a candid statement of the radical project of the chemical revolution; Lavoisier meant to eradicate all the misconceptions that had accumulated before his time and redefine the science from a point zero.

The new clarity of observation-based ideas must be expressed by a newly clarified language. Lavoisier's concluding quotation from Condillac is another description of the symbiosis of language, thought, and scientific method: "But, after all, the sciences have made progress, because philosophers have applied themselves with more attention to observe, and have communicated to their language that precision and accuracy which they have employed in their observations: In correcting their language they reason better."

THE CONTENT OF *Traité* was not especially new; rather it was an organized reiteration and reaffirmation of work that Lavoisier had previously completed and theories he had previously announced. The third and final section was mostly technical—describing the exact methods for repeating the experiments on which the new chemical theory was based. This section was elaborately illustrated by Marie-Anne Lavoisier's thirteen engraved plates depicting the instruments of his laboratory—the ocular proof that Lavoisier's chemistry was up-to-the-minute, state-of-the-art science.

Meanwhile, the opening section of *Traité élémentaire* reprised the seminal work accomplished in pneumatic chemistry during the previous decade—covering the discovery of oxygen, the composition of water, and Lavoisier's oxygen-based theory of combustion. This section also included numerous analyses of oxides and acids in terms of their bases. In a chapter on "the vinous fermentation," Lavoisier (in an apparently casual aside) stated completely for the first time in the history of science the principle of conservation of matter, which ruled his whole experimental method: "We may lay it down as an incontestible axiom, that, in all the operations of art and nature, nothing is created: an equal quantity of matter exists both before and after the experiment; the quality and quantity of the elements remain precisely the same; and nothing takes place beyond changes and modifications in the combination of these elements. Upon this principle the whole art of performing chemical experiments depends."

Before Lavoisier, the central enterprise of chemistry had been the definition of the primary elements of matter; Lavoisier, with his stress on "changes and modifications in the combination of these elements," shifted the emphasis to chemical reaction. Henceforward, all formulations of what matter *is* would be based on chemical combination.

Of course, Lavoisier had been struggling toward a comprehensive theory of the elements since the early 1770s, but his handling of the subject in *Traité élémentaire* was, at the outset, almost dismissive:

> All that can be said upon the number and nature of the elements is, in my opinion, confined to discussions entirely of a metaphysical nature. The subject only furnishes us with indefinite problems, which may be solved in a thousand different ways, not one of which, in all

probability, is consistent with nature. I shall therefore only add upon this subject that if, by the term *elements* we mean to express those simple and indivisible atoms of which matter is composed, it is extremely probable that we know nothing at all about them; but if we apply the term *elements*, or *principles of bodies*, to express our idea of the last point which analysis is capable of reaching, we must admit, as elements, all the substances into which we are capable, by any means, to reduce bodies by decomposition. Not that we are entitled to affirm, that these substances we consider as simple may not be compounded of two, or even of a greater number of principles; but, since these principles cannot be separated, or rather since we have not hitherto discovered the means of separating them, they act with regard to us as simple substances, and we ought never to suppose them compounded until experiment and observation has proved them so.

In the preface to *Traité*, this distinctly cautious passage serves as a bridge to a discussion of the new chemical nomenclature, which Lavoisier reiterated in the book's second section (as did Guyton de Morveau in his introduction to the second volume of *Encyclopédie méthodique*). Lavoisier's second section takes the form of a series of tables listing substances, with their names in the new nomenclature paired with their names in the old chemical lexicon. The elements presented in Lavoisier's "Table of Simple Substances" (with the notable exceptions of "Light" and "Caloric") are mostly retained in the modern periodic table. With all its permutations, variations, and elaborations, the Aristotelian four-element theory had now passed into history.

———

THE YEAR 1789 saw the publication of Lavoisier's *Traité élé-mentaire*, Guyton de Morveau's new volume of *Encyclopédie méthodique* with its renunciation of phlogiston and affirmation of the new chemistry, and a new edition of Fourcroy's *Éléméns d'histoire naturelle et de chimie*, which also published the new nomenclature. One year later, Chaptal's *Élémens de chimie* reaffirmed the virtue of names that made clear the chemical composition of substances (in a further clarification, he changed the inconsistent name "azote" to "nitrogen"). The establishment of *Annales de chimie* meant that these works found favorable reviews. More and more members of the scientific community laid down their phlogiston to join the chemical revolution.

Lavoisier's own conception of "revolution" was not exactly fire-breathing. What brought about his "swift and necessary revolution" was "a well-made language, a language in which we have seized the successive and natural order of ideas." Though radical in its implications, this vision of revolution was unusual for its orderliness and its composure. In his famous notebook entry of 1773, Lavoisier embraced the idea of scientific revolution with a more abandoned passion than he would show in 1789, when the chemical revolution he had instigated really amounted to the establishment of a new, incontrovertible order.

Revolution as Lavoisier saw it had the stately inevitability of planets revolving around the sun. The streets of Paris running with blood did not enter into his vision. The replacement of one order by another was revolution enough for him. He did not foresee an order succeeded by anarchy.

He had four copies of *Traité élémentaire* specially bound in red leather for presentation to the comte de Provence, the comte d'Artois, and the king and queen of France. These were ceremonial necessities, though Lavoisier took them seriously, fretting that the appearance of the volume was not so elegant as it should have been. More important to the recognition of the New French Chemistry was the presentation to Benjamin Franklin (whose response to the *Méthode de nomenclature chimique* had been no more than tepid).

In February of 1790, Lavoisier sent to Franklin in the United States two copies of *Traité élémentaire*, accompanied by an extremely revealing letter:

> *Sir and most illustrious colleague:*
>
> *M. de Coullens de Caumont, who is returning to America, has been so good as to take with him two copies of a work which I published here about a year ago under the title "Treatise on the Elements of Chemistry," one of which I beg you will accept with my best wishes, passing on the other to the "Society" in Philadelphia.*
>
> *In all the chemical treatises which have been published since Stahl, the authors have begun by laying down a hypothesis and have attempted to show that with this given principle all the phenomena of chemistry might be accounted for tolerably well.*
>
> *I believe, and a great number of scholars today agree with me, that the hypothesis advanced by Stahl, and since modified, is erroneous, that phlogiston in the sense that Stahl understood the word does not exist, and it is principally to develop my ideas on this subject that I undertook the work which I have the honor to send to you.*
>
> *I have sought, as you will observe in the preface, to*

arrive at the truth by a coordination of facts, to suppress theorizing as much as possible, which is often a false instrument and has often led us to stray from following the light of observation and experiment as closely as we might.

Here Lavoisier explicates his method and approach still more plainly and succinctly than he had done in *Traité* itself. His program was not merely to revise and replace the content of chemistry but to reverse the whole direction of thought on the subject—which would no longer flow from hypotheses toward phenomena, but from "observation and experiment" toward proven conclusions. The result:

This course, which has not hitherto been followed in chemistry, has given me an occasion to classify my work in an entirely new way, and chemistry has been brought much closer than it was to experimental physics. I sincerely hope that your leisure and your health will permit you to read the first chapter, in as much as your approval and that of those scholars of Europe who are without prejudice in matters of this sort is the object of my earnest ambition.

It seems to me that to present chemistry in this form is to render the study infinitely easier than it has been. Young men whose minds are not preoccupied with any other system seize it with avidity, but old chemists still reject it, and most of these have even more trouble in comprehending and understanding it than those who have not yet made any study of chemistry.

Lavoisier really had accomplished his pedagogical mission, to create a course in chemistry whose sensible, intelligible order would appeal to the sort of mind he had possessed in his own student days. That accomplishment was (as Lavoisier kept stressing) a strong guarantee that the new chemistry would be wholly adopted by the rising generation of young scientists. Nonetheless, the political problem of winning acceptance of the new theory from the current roster of international scientific eminences still worried him; it was here that Franklin's approval could carry a critical weight.

> *French scholars are divided at this moment between the old and the new doctrine. I have on my side M. de Morveau, M. Berthollet, M. de Fourcroy, M. de La Place, M. Monge, and in general the physicists of the Academy. The scholars of London and of England are also gradually dropping the doctrine of Stahl, but the German chemists still adhere to it strongly. This then is the revolution which has occurred in an important branch of human knowledge since your departure from Europe; I look upon this revolution as well advanced and it will be complete if you will stand with us.*

This language, and these analogies, were calculated to appeal to Franklin both as a scientist and as a politician, for Franklin had been a significant figure in the American Revolution (as no one knew better than the French). Having constituted his call for Franklin's support on a very stable, well-reasoned basis, Lavoisier shifted the subject:

> *Now that you have been informed as to what has transpired in chemistry, it might be well to speak of our political revolution. We look upon this as accomplished and accomplished irretrievably.*

So far as the chemical revolution was concerned, Lavoisier's optimism was justified. Scientists have stood with him in this matter from his day to ours. But his sense that the French Revolution was over was a very drastic miscalculation, and one that eventually cost him his life.

V

The End of the Year One

Roughly a month after the fall of the Bastille on July 14, 1789, the director of the annual Salon de Peinture requested that Lavoisier remove the elaborate portrait of the scientist and his wife that David had completed not long before, and that had only recently been installed in the exhibition. Now the management of the gallery was concerned that the image—the Lavoisier couple at the pinnacle of their professional and social success—might excite unfavorable attention from an increasingly volatile public.

The portrait was replaced in the exhibition by another David work depicting Paris and Helen—a standard, innocuous, neoclassical theme. Marie-Anne Lavoisier, though she lost much of the couple's property during the Terror, hung on to the portrait and eventually left it to her great-niece, who sold it to John D. Rockefeller in 1924.

The year 1789 had seen the consolidation of the chemical revolution by the publication of Lavoisier's *Traité élémentaire de chimie*, but in other respects the spring and summer of that year had been somewhat difficult. The removal of the David portrait, with its loud iconographic statement of Lavoisier's triumph in the reform of chemistry, was not the only symptom that the scientist and his wife were at risk of losing their popularity. Intimately involved as he was at this time in the tax system and in French national finances, Lavoisier was certainly well aware of the winds of change, though he may have misinterpreted their direction and their force.

In March of 1789, Lavoisier stood for election as a representative of the Third Estate—a political interest group so designated because its members belonged neither to the nobility nor to the clergy—in the parish of Villefrancoeur, a rural area in which he owned considerable property. He believed himself popular in the region because he had spent several years doing experiments in agricultural reform on his estate at Fréchines—work on a scale to benefit the surrounding area and intended to address the rising problem of famine in France. Also he had quite recently made large interest-free loans to the towns of Romorantin and Blois for the immediate relief of grain shortages. Though these actions were essentially those of a philanthropist, the language in which he declared his candidacy somewhat anticipated the populist rhetoric that would become full bloom in the French Revolution. Promising to "renounce all financial exemptions not shared by" the people of his constituency, he announced that "from now on there will be no financial distinction separating us; we shall all be brothers and friends." This claim had its disingenuous side; whatever exemptions he renounced, Lavoisier would remain a multimillionaire and lord of the grand manor he owned at Fréchines.

In the election at Blois on March 9, Lavoisier's opponents

attacked him, with unexpected vigor, as a noble and as a member of the General Farm. The former accusation had its irony. Lavoisier's claim to nobility was not much more convincing than his effort to reposition himself as an average member of the constituency he proposed to represent; his real place in society was with the *haute bourgeoisie*. However, Lavoisier's father had bought for him the ornamental title "Secretary to the King" at the time of his marriage—and such purchased titles were becoming points of vulnerability for the people who held them but who had no real place in the hereditary nobility and could find little help or protection there. Unlike scientific colleagues such as Trudaine de Montigny and Guyton de Morveau, Lavoisier had not attached the aristocratic "de" to his name, and though he was by no means a proletarian, his interest in constitutional reform of the French monarchy was undoubtedly sincere.

Still, the attack was vicious and successful. One of his opponents was so inflammatory that Lavoisier reported that were it not for his own "extreme prudence," he did not know "to what excess the people might have been carried against him." The risk of mob violence in the Blois Assembly was not the only narrow scrape he would endure that year.

Next, Lavoisier tried for an office representing the nobility of the same region, but the nobles of Blois were queasy at his involvement in the General Farm; the customs wall around Paris, perhaps the single most unpopular action the Farm had ever undertaken, was just then nearing completion. The frostiness of Lavoisier's reception by both the nobles and the Third Estate in these elections was a clear indicator that his position as General Farmer and noble-in-name was becoming dubious. Still, he did not resign his place in the Farm for two more years.

Rejected as a candidate, Lavoisier was invited to serve the nobles of Blois as a secretary, and in this capacity he drafted a description of an ideal (if not idealized) constitution for France. Though he was recording a synthesis of the views of the group, parts of the document had a distinctly Lavoisierian flavor. A new constitution must first and foremost guarantee the security of persons and property, while at the same extending the sacred human right of freedom to all. A specific argument for abolishing all customs duties collected within the borders of France suggests that Lavoisier well understood, by this time, the advisability of distancing himself from previous practices of the General Farm. The document recommended equality in all taxation, along with reforms of the justice, agricultural, and financial systems. It concluded by saying that the States General (the body to which representatives were being elected) must not adjourn without having, created and installed a constitution.

The rest, in the well-worn phrase, is history. On May 5, Louis XVI convened the States General at Versailles. After some six weeks of stormy debate, the Third Estate declared itself to be the National Assembly of France and called for the other two estates, the nobility and clergy, to join. When the king reacted by closing the Versailles meeting hall, the Third Estate reassembled in a somewhat impromptu fashion at the Jeu de Paume, a tennis court in the Tuileries gardens in Paris. There, on June 20, the representatives swore an oath not to dissolve their body without the approval of a constitution. Seven days later, the king capitulated, at least in name and in form, and ordered the nobility and the clergy to join the new National Assembly.

At the same time, numbers of mercenary soldiers outside Paris were increased by a royal government increasingly concerned for its own security. Within the city wall, the mood of the people, exacerbated by food shortages and nervousness inspired by the troop movements, grew ever more volatile. On July 12, Bernard-Jordan de Launay, a commander of royal troops, ordered the transfer of powder stocks from the Arsenal to the nearby Bastille, which he meant to prepare as a point of defense in case the populace went out of control. In fact, popular demonstrators had already managed to seize thousands of guns from unresisting troops at the Hôtel de Invalides and the Hôtel de Bretonvilliers, the headquarters of the General Farm. Other royal commanders were planning to enter Paris with their forces.

In the midst of this explosive situation, Lavoisier straddled the widening political fissure. His laboratory was still in the Arsenal, and because of his post in the Gunpowder Administration he shared responsibility for the movement of powder to the Bastille. On July 13, the electors of the Third Estate, now claiming to control the government, appointed him as one of several commissioners in charge of maintaining order in the area where these events were taking place. The same day, rioters began demolishing the Farm's customs wall around Paris.

Lavoisier spent much of July 14 before the National Assembly, now headquartered in the Hôtel de Ville, trying to justify the transfer of powder to the Bastille while at the same time reassuring the electors that powder would be made available for the defense of the popular interest if matters should come to that pass. The matter became moot while he was discussing it, as the Bastille's garrison of 114 men dissuaded de Launay from blowing up the powder depot and surrendered to the armed crowd of thousands outside. A total of seven prisoners were liberated.

On July 17, Louis XVI arrived in Paris from Versailles. Before the electors in the Hôtel de Ville, he pinned the new revolutionary cockade to his hat—a symbol of his acquiescence to the power of what now called itself the Constituent National Assembly. The occasion was celebratory, yet the whole country remained drastically unstable. Peasant riots broke out all over the countryside, while the mobs of Paris surged through the streets, greatly jeopardizing the security of both persons and property.

Lavoisier, who had wisely turned his back on the destruction of the Paris customs wall, was now placed in nominal charge of the demolition of the Bastille. Here, as in most matters, his respect for order, his essential conservatism, and his general good sense all came into play. The mob had begun wrecking the fortress on July 14, but the job was too large and too difficult to be completed in the first burst of enthusiasm. On July 27, Lavoisier organized a contribution of over thirty-five thousand livres (sagaciously apportioned among his various sources of income, including the Gunpowder Administration, the Discount Bank, and the General Farm) to pay a contractor to carry out the demolition of the fortress in a more orderly and much safer fashion. Various techniques for dismantling the edifice were debated in the last days of July. This matter, too, went moot while still under discussion, as workers finished the chore with pickaxes.

On August 6, Lavoisier was on the site of the Bastille, wrangling over the costs of demolition with the contractor, when the whole *poudre de traîte* episode erupted—actually in view of the boat that was about to debark, carrying the suspicious powder to Essonne. Lavoisier's escape from a summary mob execution on this occasion was narrow enough to have shaken him, though he had an unexpected capacity to remain calm in such crises. He resisted the taint that the whole misunderstanding cast on his public reputation by reading a statement justifying his conduct before both the Academy of Sciences and the local committee of the Constituent National Assembly. Yet one imagines that when two weeks later he was asked to remove the David portrait from Salon de Peinture, he did not put up much resistance.

UNTIL 1789, LAVOISIER had actually enjoyed considerable political influence, though more behind the scenes than on the public stage. His posts in the Gunpowder Administration, the Discount Bank, and even the General Farm gave him leverage on many aspects of government. In all of these spheres, he acted as a reformer, often under the influence of Physiocrats like Turgot. He believed that improved techniques would lead to better government; he believed, like many of his Enlightenment contemporaries, that application of the attitudes of rational positivism to the economic and political dimensions could succeed.

After the shocks of the summer of 1789, it appeared that the situation in France might return to stability. Reason seemed to have reasserted itself with the Declaration of the Rights of Man and of Citizen on August 28. Lavoisier would have approved this proclamation, which echoed much of the rhetoric of the document he had drafted for the nobles of Blois that spring, and which, moreover, was substantially modeled on the American Declaration of Independence of

1776. Across the Atlantic, the U.S. Constitution continued to evolve in a reasonably decorous manner. The first ten amendments, constituting the Bill of Rights, were voted into place by the U.S. Congress in September of 1789. With other moderates, Lavoisier hoped and expected that the French Constituent National Assembly would follow a similar course, arriving at a constitutional monarchy not wholly dissimilar to the one that existed in England, though more liberal in its rhetoric and perhaps even in its practice.

Such expectations failed to take into account the essential differences between the French and American Revolutions, which, at such close proximity, were admittedly difficult to discern. The American Revolution had its origins in a tax revolt and ended with the separation of the colony from the colonizing power. Despite its rhetoric of "life, liberty, and pursuit of happiness," its extension of political rights to its citizenry was quite sharply limited in the beginning, and (as the French radicals would soon begin to declaim) it left the government mostly in the hands of large Virginia landowners who depended on the labor of slaves. Moreover, it left the structure of colonial society relatively intact.

The French Revolution, by contrast, was a genuine class revolt that erupted in the heart of the nation rather than on its colonial periphery. In the beginning, it too was compromised in some respects—the French economy had a drastic dependence on slavery in its Caribbean colonies, especially the wealthy Saint Domingue, which, as the economy at home disintegrated under the pressures of increasing disorder within French borders and multiple foreign wars, would become the sole viable source of revenue. Yet the French Revolution always had its radical wing, which would accept no compromises and was determined to follow the ideology of liberation to its limit, regardless of economic or any other consequences. When the radicals came to power, all aspects of the old French society would be uprooted and overturned.

Like most of his fellow moderates, Lavoisier was slow to recognize this crucial distinction. But another shock came in early October of 1789. Food riots, instigated by a chronic shortage of flour for bread, had become common in Paris. On October 5, Marie-Anne Lavoisier (devoted pedestrian that she was) got her carriage caught in one of those mobs of angry women whose excesses had begun to surpass those of the men. She was forced to get down and walk among these dangerous *tricoteuses* until, presumably, they tired of tormenting her and left her free to join her husband and the American Gouverneur Morris for a luncheon planned at the Arsenal apartment. In fact, the *tricoteuses* had larger game in view; they were en route to Versaille to capture "the baker, the baker's wife and the baker's boy"—their sarcastic phrase for the royal family. On October 6, after much indignity, the king, queen, and dauphin were brought back to Paris and to all intents and purposes imprisoned in the Tuileries. The National Assembly set up shop in an adjoining hall. In the wake of this episode, Lavoisier began to do what he could to detach himself from politics, though he remained active in the Discount Bank for a couple more years, and also found it difficult to avoid exposure to the scandals provoked by the constant anxiety about munitions.

EVER HOSTILE TO the French Revolution, the English commentator Edmund Burke described it thus: "The wild *gas*, the fixed air, is plainly broke loose: but we ought to suspend our judgment until the first effervescence is a little subsided, till the liquor is cleared, and until we see something deeper than the agitation of a troubled and frothy surface." Burke chose this language to attack Lavoisier's old rival Priestley, who was himself a revolutionary sympathizer. Lavoisier might have enjoyed the analogy, and endorsed the idea of suspending judgment; he was eager for the revolutionary "effervescence" to subside.

When he wrote to Benjamin Franklin on February 2, 1790, he wanted to believe that it had subsided: "We regard [the political revolution] as complete and complete beyond return; however there exists an aristocratic party, which makes vain efforts and which is obviously the weaker. On the other hand, the democratic party is most numerous and moreover has education, philosophy and the Lumières on its side. Moderate persons, and those who have kept their cool during this general effervescence, think that circumstances have taken us too far, that it is unfortunate that we have been obliged to arm the people and all citizens; that it is impolitic to place force into the hands of those who ought to obey; and that it is to be feared that the establishment of the new constitution will beget obstacles on the part of those same people for whose benefit it has been made."

In describing the position he thought of as moderate, Lavoisier betrays some nostalgia for the *ancien régime*; in stating that the French Revolution has passed the point of no return, he clearly also wishes that elements of it could be turned back. No doubt many moderates shared his nervous mistrust of the extremes that revolutionary populism might reach. No doubt he was absolutely sincere in this wistful sentence to Franklin: "At this moment we regret very much your distance from France; you would have been our guide, and you would have marked for us the limits which we should not have exceeded."

Between these lines one reads that while Lavoisier wished very much that the French Revolution was over, he was too perspicacious not to suspect very strongly that there would be more and worse to come. Yet it does seem, from the course that he followed from 1790 to 1793, that wishful thinking did something to impair his perception. Lavoisier was always well out in front of the curve of the chemical revolution, but in the case of the political revolution, he lagged a little behind.

MARAT, WHOSE SCIENTIFIC nemesis Lavoisier had been, published the first edition of his newspaper *L'ami du peuple* in September of 1789. At last he had found substantial "matter of fire"—in the form of incendiary journalism. He would devote these last few years of his life to goading and whipping up the radical left, but he always had a moment to spare for old enemies; unmasked by Lavoisier as a scientific charlatan, he could hit back by denouncing Lavoisier as "the Coryphaeus of the charlatans." By 1791, the old order had been so thoroughly inverted that Marat could cast Lavoisier in a highly unfavorable light by merely reciting his resumé: "Farmer General, Director of the Gunpowder Administration, administrator of the Discount Bank, member of the Academy of Sciences, Secretary to the King," and so on.

Jacques-Pierre Brissot, who had been early to accuse the national academies of being intrinsically tyrannical and who had been stung by Lavoisier's debunking of the mesmerism cult, joined the hounding with an article in *Le patriote Français*: "General Farmer and Academician, two titles for the encouragement of despotism, still worse he is the author of the plan to build a wall around Paris. Shouted down at Blois when he presented himself for election, the only votes he got were for charity. Lavoisier became a chemist; he would have become an alchemist if he had followed only his unquenchable thirst for gold."

Lavoisier made no reply to these attacks. He had learned that silence might be the best means of preserving his dignity, since he had made the miscalculation, a short time before, of publishing that open letter in *Le moniteur* announcing that in the spirit of brotherhood and equality he would decline salaries for all of his five public posts, excepting the Gunpowder Administration. The contortion resembled the one he had tried in making himself "equal," under the tax code, to the constituency of Villefrancoeur, and the outcome was no more successful. A deafening backfire was immediately heard from the far-left journal *Les actes des apôtres*, in the form of a satirical poem:

> *Généreux Lavoisier, ta lettre pathétique*
> *M'a fait, je l'avouerai, presque verser des pleurs;*
> *Tu viens de conquerir à la fois tous les coeurs*
> *En nous développant ta conduite héroïque.*
> *Quel exemple étonnant de modération!*
> *Se contenter d'avoir cent mille écus de rente*
> *Qu'on a gagnés, Dieu sait! puis à sa nation,*
> *Donner son temps pour rien, cet exemple me tente.*
> *Ah! Que ne puis-je, hélas! pouvoir en faire autant!**

---

* Generous Lavoisier, your pathetic letter
almost made me (I admit it) spill tears;
You have just conquered everyone's hearts at one stroke
In revealing to us your heroic conduct.

The not-altogether-unreasonable point that if important public posts went uncompensated, only the wealthy could afford to occupy them was echoed by Brissot in *Le patriote Français*. Meanwhile, the official response to Lavoisier's sally was not much more encouraging. He wanted above all to keep his post in the Gunpowder Administration, for the security of his laboratories, still housed in the Arsenal. He wrote frankly to Louis Hardouin Tarbé, the minister charged with that decision, underlining that "there I have established myself at my own expense, [there] I have made considerable expenditures in laboratories, in chambers related to the sciences which I pursue, and in instruments." In the end, Lavoisier agreed to give up his official post in the Gunpowder Administration on Tarbé's assurance that he could continue to occupy his Arsenal laboratories; his position became that much more precarious.

JEAN-PAUL MARAT suffered from eczema and took long soaks to relieve the condition. He was in the habit of seeing visitors while lying in a covered tub. On July 13, 1793, he unwisely agreed to receive a stranger, a young woman named Charlotte Corday; she stabbed him, fatally, and fled. The painter David rendered the death scene in a work far different in style and feeling from his portrait of the Lavoisier couple— as disagreeably realistic as a photograph. The turbaned Marat,

---

What a blinding example of moderation!
To content yourself with a hundred thousand ecus in income
Which you've earned, God knows! then to your nation
Give your time for nothing, this example tempts me.
Ah! how am I unable, alas! to be able to do as much!

his skin sallow from his illness or from loss of blood, lies slack in the coffin-like bathtub, one dangling arm still holding his quill pen, the wound in his chest a bloodless pucker. Thoroughly unprepossessing, this grim portrait is also one of David's least forgettable works, and in the end an appropriate icon for the events that surrounded Marat's demise.

For Lavoisier, the death of his enemy brought no relief. The assassinated Marat was to all intents and purposes canonized by the Jacobin Party whose champion he had been, and which was now moving into ascendancy in the national government. As a member of the National Guard, Lavoisier was required to march in a military parade to honor Marat's corpse, decked out by David in Roman costume for the occasion. A few months later he was again obliged to turn out for a similarly formal salutation of Marat's bust, unveiled on October 15. Somehow Marat had become more powerful in death, as if his ferocious spirit, released from the body, now possessed the rulers of the nation.

THE PRELUDE TO Marat's assassination was a fresh revolutionary spasm in the National Convention. This body, which replaced the legislative assembly on September 20, 1792, had for two years been divided between two factions: the Montagnards of the radical left and the more moderate Girondins. In early June of 1793, the Montagnards, led by Marat himself, Georges Danton, and especially Maximilien Robespierre, succeeded in purging the Girondins from the Convention; those who escaped arrest took flight. At the end of the month the assembly adopted the Constitution of the Year One, which transferred power wholesale to this legislature and to its executive committees, especially the notorious Committee of Public Safety, now in the hands of Robespierre.

Though the motives of Charlotte Corday were tortured and obscure, the slaying of Marat was perceived by the newly installed Jacobin Republic as a reaction from the center and the right. Doctor Guillotin's efficient death machine was already installed on the Place de la Révolution; Charlotte Corday died there on July 17, a mere four days after killing Marat. Subsequent trials would be still more speedy. To quell any further reaction, to secure its own position, and to intimidate foreign enemies and their agents within France, the Jacobin Club called for a Reign of Terror on August 30, 1793. With the passage of the "Law of Suspects" on September 17, the Terror became official.

SINCE THE MOB brought him back to Paris in October 1789, Louis XVI had been progressively disempowered, and his family's situation had become ever more fragile. Nominally a constitutional monarchy existed, or was under construction at least, until September of 1792, but the king proved increasingly unable to manage the situation through edicts and vetos, and was tempted by the possibility of reversing the Revolution altogether. The monarchical powers of Europe were on his side, and in June of 1791, the king attempted to take his family out of France to join them, but they were arrested at Varennes and again dragged back to Paris. Pressure to abandon the notion of constitutional monarchy in favor of a republic came to a head on July 17, at the celebration of the second anniversary of the fall of the Bastille. During disrupted public ceremonies on the Champ de Mars, the marquis de Lafayette, who though beloved for his role in the American Revolution was of a more conservative bent than most of his fellow French citizens, ordered troops of the National Guard to fire on the rioting republican crowd.

These events gave the left reason to fear that the revolution might indeed be reversed, by the action of royalist forces both in and outside of France. Nevertheless, a constitution was proclaimed in September of 1791, and Louis XVI was restored to constitutionally limited power. But the relationship between king and legislature soon broke down. In early June of 1792, the king dismissed the Jacobin legislators; Lafayette supported this action, but the Paris mob invaded the Tuileries, forcing the king to put on a red liberty cap and drink a toast to the sovereign French people. Robespierre was encouraged to call openly for the king's removal in a meeting of the Jacobin Club. On August 10, the Tuileries was sacked and looted by a crowd that announced its intention to bring down the monarchy. The king and his family became prisoners of the nation. On September 21, the monarchy was abolished and the French Republic declared.

Louis XVI went to the guillotine on January 21, 1793. Since the spring of 1792 France had been at war with Austria and Prussia (the queen, Marie-Antoinette, was a member of the Austrian royal family). In February of 1793, France declared war on England and Holland as well; in March, Spain was added to the list. Resentment of a new military draft required for the war effort fueled royalist revolts in Brittany and the Vendée. On March 10, the Convention created a Revolutionary Tribunal for the judgment of any and all enemies of the republic. This body was augmented on April 6 by the Committee of Public Safety, endowed by the Convention with an explicit mission to terrorize the enemies of France. The death of Marat was the last straw needed to tip the balance wholly in favor of the Terror.

WITHIN WEEKS, THE Terror had mutated into a cancer attacking the body it was supposed to protect. In the end, Robespierre himself would die of it. Lavoisier, meanwhile, like most other members of the drastically destabilized bourgeoisie, was doing all he could to avoid entanglement in the rogue machinery. He hoped, and it was not an unreasonable hope, that his status as a scientist and the material value that his prowess in the sciences could offer the republic would shield him from accusations that might stem from his other activities in finance, the Gunpowder Administration, and, most dangerously, the General Farm. But the intellectual organizations and structures that Lavoisier thought would be his refuge were already being dismantled.

From the first beginnings of his career, Lavoisier had venerated the national academies as "little republics"—miniature realms within the French kingdom that made intellectual, scientific, and artistic pursuits secure from pressures outside their borders. Even at that early date, men like Marat and Brissot (who had bloomed, then withered, as a leader of the Girondins) had objected that the academies were inherently elitist and even tyrannical. In August of 1793, the latter point of view saw its triumph. The charge against the Academy of Painting and Sculpture was led by its most distinguished member, the painter David, who addressed the Convention in a passion of implied self-criticism that would have pleased Pol Pot as much as it presumably did David's close friend Robespierre: "Let them be closed forever, these schools of flattery and servility. . . . To speak of one Academy is to speak of them all; there is the same spirit in all of them, and in all are found the same men. . . . In the name of humanity, in the name of justice, and above all for your love of youth, let us destroy—let us annihilate—these deadly academies, which can no longer survive under a free regime. Academician though I am, I have done my duty."

On the same occasion, Abbé Grégoire presented a more measured critique of the academies, with a special argument for the preservation of the Academy of Sciences, which continued, as he pointed out, to serve the nation with projects such as the new system of weights and measures. In his preroration, Grégoire underlined the astronomical achievements of Johannes Kepler, "the great mass of knowledge which we owe to the genius of Newton," and finally, "the most sublime research in chemistry." His eloquence accomplished little. The Academy of Sciences was abolished along with the rest, leaving Grégoire gloomily predicting a "doleful future" featuring "the persecution of savants."

BUT THE SAVANTS—some of them anyway—had invented the Terror themselves. Though the French Revolution had its episodes of real anarchy, the Terror was not one of them. On the contrary, it was a force for order, though not a style of order Lavoisier would have endorsed. The very senselessness of its final outcome was the product of the extension of its logic to an ultimate extreme. And the Terror *was* logical, above all.

The twenty-first-century commentator Charles Murray has gone so far as to suggest that Newton, or rather the eighteenth century's *fin de siècle* compulsion to Newtonize absolutely everything, was to blame for it all. "Reason was the new faith.

Its first political offspring was the grotesque Jacobin Republic set up after the French Revolution." It proved politically foolhardy to assume that all phenomena, including human behavior just as much as the behavior of matter in physics and chemistry, must ultimately yield to reason. "Newton worshipers," Murray argues, "decided that what could be known of the motion of bodies could be known as well of humans. Man could remake the world from scratch by designing new human institutions through the application of scientific reason."

As the Enlightenment darkens into Terror, much of the surrounding cultural context does support this view. The deistic revision of Christianity, espoused by Benjamin Franklin among others, proposes that all the universe is a vast machine, originally created by God but capable of being maintained and even upgraded by human beings. Condillac's *Logique*, which Lavoisier so admired, implied an extension of the synergy of thought, speech, and writing to the realm of action. If chemistry could be made to follow the rules of algebra, why not politics? Of course, Lavoisier would not have had the rationalization of politics go nearly so far as it did. Like most of the other embattled moderates, he preferred that it stop where he believed Franklin would have stopped it, with a more benignly humane view of politics that (in Murray's phrase) "allowed for an intractable and problematic human nature."

Condorcet had also been a moderate; he was purged from the legislature as a Girondin, then tried and condemned *in absentia* in October of 1793. His crime, in essence, was to have foreseen and protested the worst consequences of the Constitution on which the Jacobin Republic was founded in June of that same year. Given the circumstances of the author (in hiding from his former colleagues, now state terrorists), the tone of unflagging optimism expressed by Condorcet's *Sketch for a Historical Picture of the Progress of the Human Mind* seems downright bizarre, as it flies in the face of facts that he was hunted for having recognized.

Yet Condorcet still believed that the synergy of thought, speech, and rhetoric could and would be unerringly translated into action and even into political action. Still he believed (with Robespierre and other members of the Committee of Public Safety) in "the true perfection of mankind." In the conclusion of the work he asks what was clearly for him a rhetorical question: "The sole foundation for belief in the natural sciences is this idea, that the general laws directing the phenomena of the universe, known or unknown, are necessary and constant. Why should this principle be less true for the development of the intellectual and moral faculties of man than for the other operations of nature?"

Material progress in physics and chemistry (once an adherent of phlogiston, Condorcet had by this time recognized the superiority of the new chemistry, which he describes earlier in the *Sketch*) must certainly imply a similar pattern of progress in all human affairs, even in the political realm. Shortly before his death in a Jacobin prison, Condorcet happily predicted, "The time will therefore come when the sun will shine only on free men who know no other master but their reason; when tyrants and slaves, priests and their stupid or hypocritical instruments will exist only in works of history or on the stage; and when we shall think of them only to pity their victims and their dupes; to maintain ourselves in a state of vigilance by thinking on their excesses; and to learn how to recognize and so to destroy, by force of reason, the first seeds of tyranny and superstition, should they ever dare to reappear among us." Of course, by the time Condorcet finished his monograph, that state of vigilance had itself evolved into tyranny.

At the time of his writing, priests were already history. Toward the end of 1789, the property of the Catholic Church was expropriated by the state. (Lavoisier, perhaps unwisely, bought up considerable acreage of former Church lands in the region of his ancestral village, Villers-Cotterêts) In 1790, all monastic and most other religious orders were abolished. In October of 1793 the new Revolutionary Calendar was adopted—rationalizing the measurement of time and detaching the system from the clutch of the Christian calendars, with their cycle of feasts, penances, and saints' days. A month later, Robespierre, who required an outlet for his piety as much as any other puritan, organized a "Festival of Reason," for 20 prairial.

Celebrated in the erstwhile cathedral of Notre Dame, this fête was designed to supplant the uprooted symbolism of the Catholic Church with a new iconography based entirely on secular principles. In keeping with his bureaucratic turn of mind, Robespierre expressed "the Cult of the Supreme Being" as a list of fifteen edicts. Article VII declared as subjects of celebration "the Supreme Being, and Nature; the human race; the French equality; the benefactors of mankind; the martyrs of freedom; liberty and equality; the Republic; the liberty of the world; patriotism; hatred of tyrants and traitors; truth; justice; modesty; glory and immortality; friendship; temperance; courage; good faith; heroism; impartiality; Stoicism; love; conjugal fidelity; fatherly affection; mother-love; filial piety; childhood; youth; manhood; old age; misfortune; agriculture; industry; our ancestors; posterity; happiness." When twentieth-century Christian fundamentalists railed against "secular humanism," this sort of thing is what they had some-where in mind.

Robespierre devoted what would have been Christmas Day of 1793 (if December 25 had still existed) to a discourse on the difference between constitutional government, which the French Revolution had been meant to establish, and the revo-lutionary government now necessitated by the exigencies of war. "The principal concern of constitutional government is civil liberty; that of revolutionary government, public liberty. Under a constitutional government little more is required than to protect the individual against abuses by the state, whereas revolutionary government is obliged to defend the state itself against the factions that assail it from every side." In this risky situation, Robespierre saw a crying need to "affright the enemies of France." And so: "To good citizens revolution-ary government owes the full protection of the state; to the enemies of the people it owes only death." The logic appeared to be inexorable; it had the Committee of Public Safety behind it in case anyone presumed to think otherwise.

Even today, Robespierre and his cohort have their defend-ers, who argue that the Terror was effective in unifying France and ensuring its survival and ultimate triumph at a time when the nation was indeed at war with practically all the rest of Europe. It did so at the cost of sending an estimated eighteen thousand French citizens to the guillotine and of creating the blueprint for systematic state terror employed by most totali-tarian governments ever since. The pseudoscientific, pseudo-algebraic methodology of the Terror resurfaces in Joseph Stalin's misappropriation and misapplication of the theories of Karl Marx—to the tune of a much, much larger death toll, exacted behind the screen of politics and rhetoric that would become known as *Orwellian*. That same methodology injects its banality into evil during the Nazi regime, by allowing men like Adolph Eichmann to practice genocide as the application of a scientific and utilitarian technique. We taste it again in the muddy, bloody killing fields of Cambodia, where Pol Pot, more faithful to the French model than most, thought to start history all over again from scratch with his very own Year One.

LAVOISIER BIOGRAPHER JEAN-PIERRE POIRIER identifies seven scientists slain by the Terror. An honorary Academician, the duc de La Rochefoucauld d'Enville, protested mob justice in general and the Revolutionary Tribunal in particular; he was killed by a mob in September of 1792. Astronomer Jean-Silvain Bailly, a political ally of Lafayette, was condemned by the Revolutionary Tribunal in November of 1793. The baron de Dietrich, a chemist in whose home the revolutionary anthem *La Marseillaise* had been composed, took a royalist part and went to the guillotine in December of 1793. Condorcet, as we have seen, died of uncertain causes in prison in April of 1794. That same month saw the decapitation of Bochart de Saron, an astronomer and mathematician who had signed a protest against the suppression of parlement, and Lamoignon de Malesherbes, who had taken part in the defense of King Louis XVI.

Yet most members of the dismantled French scientific establishment managed to keep their heads and their lives—including several of Lavoisier's closest colleagues: Meusnier, Monge, Fourcroy, Berthollet, and Guyton de Morveau. Even in its terroristic frenzy, the Jacobin Republic did not completely lose sight of the value of scientific skills to the nation, so several of these men were still honored in the public eye. The rest relied on discretion, on avoiding notice of any kind. All the survivors remained in France for the duration. The only scientist who attempted flight, the baron de Dietrich, was finally condemned as an *émigré*.

AS FOR LAVOISIER himself, he seems to have found it difficult to comprehend how completely the world around him had lost its bearings. He (who had always trusted his scales) firmly believed in *balance* in all things, and so believed that equilibrium must certainly restore itself. He had a project to spend some months in Scotland, collaborating with Joseph Black, and hindsight shows that he would have been better off to have carried it out in the fall of 1793. Yet such a move might have caused him to be condemned as an *émigré*, along with the baron de Dietrich and hundreds of others. He may have been held in France by concern for his laboratory and for his family, perhaps also by pride, stubbornness, and the deep reserve of courage he would prove in his last days.

Moreover, he could reasonably suppose that he might ride out the storm. After all, he was probably the most prominent among the coterie of scientists whose status did ultimately protect them, even after the academies had been abolished. And in his multiple, wide-ranging services to the nation, he had been more diligent, more energetic, and more successful than almost any of them. It was simply unbelievable that the French Republic would be so profligate as to throw away his talents.

Nonetheless, his papers were seized and sealed just three days after the adoption of the Law of Suspects. Lavoisier might have gotten out of France then, as no move had been made to arrest him, but he elected to stay and to struggle. At first his efforts seemed successful. On September 28, seals were removed from the Farmers' papers and Lavoisier was awarded a "certificate of civic virtue," noting that everything in the suspect documents "does honor to your civic spirit and is susceptible to dissipate any sort of suspicion." With that, Fourcroy quietly returned the correspondence that had been confiscated from the Lavoisier residence on boulevard de la Madeleine.

The wording of the certificate seemed the best insurance possible against the Law of Suspects, so Lavoisier might be forgiven for thinking himself safe. In fact, he had merely gained a tempo in an endgame that had already begun. On November 24, an order went out for the arrest of all former shareholders in the General Farm, who were to be imprisoned until the suspect accounts of the Farm should be rendered and judged. The "certificate of civic virtue" was no longer worth whatever paper it was written on.

The police missed Lavoisier at the Arsenal laboratory and the house on boulevard de la Madeleine; they were unaware that the chemist was devoting that day to service with the National Guard. For three more days, Lavoisier remained at large, writing feverish letters to the Committee on Public Education and to the Committee of General Security—describing himself in the third person, as if to define their attitude toward him in advance.

The Committee on Public Education was in charge of the weights and measures project on which Lavoisier was still working. Moreover, Lavoisier had spent much of the previous year developing projects for the reform of education, culminating in a report supporting the Lycée des Arts sent to the

Convention in July of 1793. His interest in the sciences had
always been entwined with an interest in pedagogy, and he
clearly saw the practical importance of sound training in the
arts and trades (whose traditional guilds had been abolished
by the Revolution some time before the suppression of the
academies). "It is not with celebrations that the United States
of America have become a flourishing nation," he wrote in the
summer of 1793. "It is by giving their industry all the development it can absorb." In this area, if in no other, he might
continue to be of significant service to the nation.

His November 25 letter to the Committee on Public Education was artfully phrased:

> *Lavoisier, of the former Academy of Sciences, quit the Gen-*
> *eral Farm about three years ago. . . . It is public knowledge*
> *that he never involved himself in the general affairs of the*
> *farm, which were conducted by a small committee named*
> *by the minister, and moreover the works he has published**
> *attest that he has always principally occupied himself with*
> *the sciences. He was never among the commissioners*
> *named in the decrees for the rendering of the General*
> *Farm's accounts; therefore he cannot be held responsible*
> *for the delay for which that commission is reproached; he*
> *does not believe that he can be included in the order that*
> *the General Farmers be placed under arrest until the ren-*
> *dering of their accounts.*

This formula, if the Committee on Public Education had
chosen to make use of it, would have neatly separated Lavoisier

---

* Here Lavoisier meant to remind the committee of the value of his
Méthode de nomenclature chimique and the *Traité élémentaire*.

from his dangerous association with the Farm, and Lavoisier
made sure to furnish a constructive reason for doing so:

> *In a state of doubt, he begs the the National Convention to
> let him know if its intention is that he apply himself to the
> accounts of the General Farm, a work to which he believes
> himself to be ill-suited, or if he should continue to fulfill his
> functions in the Commission for Weights and Measures,
> for which he has worked up to now with zeal and, he dares
> to say, with some usefulness.*

On November 26 he wrote in a similar vein to the Committee of General Security—a body which despite the similarity
of title (or because of it) tended to be opposed to the Committee of Public Safety, now dominated by Robespierre and
most ardent terrorists. Lavoisier had been eluding arrest for
some forty-eight hours, and in this second letter he
announced a willingness to give himself up:

> *Lavoisier, of the former Academy of Sciences, is charged by
> the decrees of the National Convention to work toward the
> establishment of new measures adopted by the National
> Convention. On the other hand, a newly rendered decree
> orders that the General Farmers be confined in a house of
> arrest to work toward the rendering of their accounts. He
> is prepared to turn himself in there, but he believes that
> beforehand he should ask which one of these decrees he is
> supposed to obey.*

So put, the question seemed a fair one, and a fair balance of
interests should have left Lavoisier at liberty, or so he must

have believed. He went so far as to suggest a compromise—that he be placed under the guard of "two of his *sans-culotte* brothers," a solution that presumably would have left him freedom of escorted movement around Paris, enough mobility to continue his scientific work.

November 27 passed with no reply from either committee. Lavoisier was in a quandary; he might still have fled the country, or gone into hiding for the long term as Condorcet had done. It was not from sheer folly that he followed neither course. Jacques Paulze, his father-in-law, was tangled in the same net with the rest of the General Farmers, and apparently it was not an option for the entire family to take flight, though Paulze was also still at large during those tense November days. And Lavoisier's picture of the future was insufficiently pessimistic in some respects. He was trained in the law, though he had never practiced, and he and Paulze shared a confidence that a fair trial would ultimately vindicate the Farmers. Lavoisier was enough of a pragmatist to have realized that his vast wealth was a target, but at this stage of the Terror he did not recognize that the state might also have designs on his life. The scientists who had thus far gone to the guillotine had been involved in explicitly political actions that Lavoisier had been very careful to eschew. In his own worst-case scenario, he would accept the confiscation of his fortune and afterward earn his livelihood as a pharmacist, living modestly, like the Swedish chemist Scheele. Lavoisier's regard for order was very profound; an inevitable return to equilibrium would surely leave him on his feet. And in the end, he could not find it in his character to run.

On November 28, Lavoisier and Paulze (neither a very frightening fugitive) surrendered at Paris's Port Libre Prison, where the other General Farmers were already incarcerated. There they remained for most of the next month. Lavoisier still hoped that a reply to one of his late-November letters might release him, and similar petitions were made by his friends on his behalf throughout December—none were successful. As for the chances of a fair trial, another captive farmer, Etienne-Marie Delahante, had a clearer, grimmer view: "I foresaw that the commissioners would accuse us of fictional abuses, that we would not be allowed to defend ourselves against the charges, and would be judged guilty of these alleged corrupt practices; thus we would be doomed."

In fact, the leaders of the Terror had Occam's razor firmly in their grasp. The greatest threat to the Farmers' survival was the voracity of the wartime economy. The Farmers were possibly the richest private capitalists in all France. During the five months of their imprisonment, the situation devolved to the point that the guillotining of the rich became jocularly known as "minting coins on the Place de la Révolution."

IN THE PORT LIBRE PRISON, the Farmers did not even have access to the accounts they were required to present. Finally on December 25, 1793 (while Robespierre explicated the distinctions between constitutional and revolutionary government), they were transported to the Farm's former headquarters in the rue de Grenelle St-Honoré, there to be confined among their own records. Characteristically, Lavoisier threw himself into the work to which he had declared himself to be ill-suited. By the end of January the accounting was done.

As the affairs of the Farm were tangled and tedious, the defense against the charges was legalistic and tedious. Furthermore, Delahante's fears were borne out by the fact that the specific charges were never directly presented to the accused Farmers, who had to guess at them on the basis of hearsay from family and friends. The difficulty of preparing a defense under such circumstances can be imagined. Nevertheless, Lavoisier did manage to prepare precise responses, though biographer Poirier notes that "they had a technocratic character which even today makes them very hard to understand." By May of 1794, given the leveling forces at work, the membership of the Revolutionary Tribunal was in no mood for the subtleties of financial statements, or for the niceties of legal procedure—or, indeed, to hear reason of any kind.

Since September of 1793, one Antoine Dupin had been charged with the investigation of the Farm's affairs. He was himself a former employee of the tax corporation, and had been until fairly recently imprisoned for stealing a significant sum from the Farm and forging the books to cover it. But Dupin was a small fish with a line on a much bigger one. He freed and empowered himself by representing himself as a victim of the Farm's general nefariousness and by claiming a knowledge of the Farm's dishonest accounting. As Delahante had predicted, Dupin's charges were distorted wherever they were not entirely fictitious.

Dupin spent nearly three months preparing his own prejudicial reports on the Farm's accounting. During that time, Marie-Anne Lavoisier managed to reach him through intermediaries, who worked out an arrangement that Dupin would treat Lavoisier's case separately (as Lavoisier had all along been trying to engineer), that Dupin would report less severely on Lavoisier than on the other Farmers, and that an opportunity might even be created for him to escape. Supposedly, all Madame Lavoisier needed to do was to call on Dupin personally, to thank him and to confirm the deal.

When she needed to humble herself, Marie-Anne instead stood on her pride, and indeed appears to have lost her temper. During her meeting with Dupin she insisted that her husband was innocent (as indeed he was) and abused his accusers for corruption and villainy (as they, including Dupin, most likely deserved).* Her comportment was just, but improvident. Of course, her father was not to be included in the special deal that might have saved Lavoisier, so to accuse the accusers was the only way she could defend them both. Whatever her motives, her outburst completely destroyed the diplomatic bridge to Dupin.

Many of Lavoisier's fellow scientists helped lobby for his release in the first months of his imprisonment, but the more it appeared that he would be unable to disentangle himself from the Farm's predicament (and the more extreme the Terror became), the more they were inclined to fall silent. The behavior of Fourcroy, in particular, has been much examined and criticized. At the time, a Dr. Sacombe denounced him: "If a man is so cowardly as to keep silent when, with a single word, he could save a great man's life, then he ought to know at least how to expiate his silence."

Fourcroy had been Lavoisier's protégé in the beginning, but by 1794 his position was much more secure than that of his mentor in chemistry; he was a member of both the Convention and the Jacobin Club. His detractors suppose that he

---

* Later on, Madame Lavoisier emerged as a leader of the survivors of the General Farmers who brought Dupin to trial and conviction in 1795.

abandoned Lavoisier from professional jealousy and a desire to supplant the senior chemist. Such behavior was not unheard of; Lavoisier in his own prime had demolished his share of seniors and rivals. But, in fact, Fourcroy tried a number of stratagems to protect Lavoisier, and shortly before the Farmers were brought before the Revolutionary Tribunal, he went so far as to crash a meeting of the Committee of Public Safety. In the description of another contemporary, André Laugier, "He spoke in favor of Lavoisier; he explained, with the heat which was natural to him, what a terrible loss this great chemist would be to the sciences. Since Robespierre, then President of the Committee, said nothing, no one else dared to reply, and M. de Fourcroy was obliged to depart without anyone having seemed to pay the least attention to what he had said. No sooner was he out of the room than the President complained of his audacity and made threats against M. de Fourcroy which terrified Prieur de la Côte D'Or to the point that he ran after Fourcroy and told him not to try it again if he wanted to save his own head."

OVER THE FIRST few days of May in 1794, Dupin presented an extremely pejorative report on the Farm's accounts, in no way excepting Lavoisier's share in them, to the National Convention. On May 5, the Farmers (still confined in their former headquarters) were notified that the Convention had voted to send them before the Revolutionary Tribunal. Though technically a trial would take place, everyone understood that this vote amounted to a death sentence. Two men who had procured opium for a suicide attempt offered to share their opportunity with Lavoisier. His response, whose relentlessly pragmatic tone must have been somewhat maddening to his auditors, still would have done honor to the Stoics: "I don't cling to life any more than you; I have made the sacrifice of my life. The last moments we await will doubtless be painful, but we should not be sure of preventing them by the means you propose; suffocation would serve us better. But why go running ahead to meet death? Is it because it is shameful to receive it on the order of another, and above all, by an unjust order? But here, the excess of injustice itself erases the shame."

On May 6, the group was transferred to the Conciergerie, a medieval fortress on the Île de la Cité, where Marie-Antoinette had spent her last days before execution. The following day, they were briefly interrogated. That evening, they were given for the first time a written statement of the charges against them, but were ordered to put out their lights before they could read these documents; they were not able to know the exact accusations before dawn of the next day.

It hardly mattered. The trial, which began at ten the next morning, was the sort of kangaroo court Delahante had anticipated. Though the remark that "the Revolution has no need for scientists" is considered to be apocryphal, the judges did amuse themselves by mocking the defendants' responses. Condemned, they were taken to the Place de la Révolution in open carts along the Seine, a route that passed under the windows of the Louvre and through the scenes of Lavoisier's best triumphs. One of the doomed Farmers, Papillon d'Auteroche, glanced at the abusive crowds that chased the carts and muttered, "What galls me is to have such unpleasant heirs."

By five that evening, the condemned men arrived at the

foot of the guillotine. A witness named Eugène Cheverny wrote afterward that Lavoisier "prepared the others for death," but in what exact terms he did not record.

THE FARMERS WERE all "decent men," Cheverny also wrote. France was losing stomach for the Terror. In less than three months, Robespierre would be dead. The day after Lavoisier's execution, Joseph-Louis Lagrange, an ex-Academician who had recently collaborated with Lavoisier on the Advisory Board for Arts and Trades, felt bold enough to say, "It took them no more than a moment to make that head fall, and a hundred years may not be enough to produce another one like it."

THOUGH THE LAST known portrait of Lavoisier is not believed to have been drawn from life, it is nonetheless impressive. A full profile engraving by Marie Renée Geneviève Brossard de Beaulieu, it shows the chemist with his hair loosened and his collar laid open for the guillotine's blade—the conventional style for such foot-of-the-scaffold images. Lavoisier's features look sharper and harsher than in the more comfortable portrait by David—understandably, given the privation and stress he had endured for the last several months. He appears calm, confident, slightly imperious, completely in command of himself, a little contemptuous (as always) of his intellectual inferiors, and absolutely contemptuous of the stupidity of what is about to happen to him.

In the Port Libre Prison, when he still had some hope, he had already begun the work of resignation. In December he wrote to Marie-Anne, "You give yourself, my dear good friend, much pain, much fatigue of body and spirit, and I—I cannot share it. Take care that your health is not altered; that would be greatest of misfortunes. My career is well-advanced; I have enjoyed a happy existence ever since you have known me. You have played and still play a great part in it every day by the marks of attachment you give me; finally, I shall leave behind me always memories of consideration and esteem; so my task is done, but you, who still have the right to hope for a long existence, don't be prodigal with it. I believe I noticed yesterday that you were sad. Why should you be, since I am resigned to everything and since I regard everything I will not lose as won?"

Typically, Lavoisier switched gears and subjects before emotion could overtake him, finishing the letter with instructions to his wife as to how and when to negotiate some banknotes.

On May 6, when informed of the Convention's vote against them, the other Farmers turned to writing their immediate families. By contrast, Lavoisier's last surviving letter is to no close intimate but to a cousin, Auger de Villers, who does not otherwise much figure in his story:

I have had a reasonably long and especially happy career, and I believe that my memory will be accompanied by some regrets, perhaps even by some glory. What more than that could I have desired? The events in which I find myself enveloped will probably spare me the inconveniences of old age. I shall die whole—still another advantage among all I have enjoyed. If I do suffer some painful feelings, it's for not being able to do more for my family, for being stripped of everything and so being unable to give either them or you any token of my attachment and my gratitude.

So, it is true that the exercise of all the social virtues, important services rendered to the nation, a career always

spent on the progress of the arts and of human knowledge
are not sufficient to prevent a sinister end or to avoid per-
ishing as though one were guilty of something.

I write you today, for perhaps tomorrow I will no longer
be permitted to do it, and because it is a sweet consolation
for me, in these last moments, to fill my mind with you and
other people who are dear to me. Remember me to those
who are interested; let this letter be shared among them. It
is most likely the last one I will write you.

If he had directly addressed his "dear good friend" Marie-
Anne, perhaps he might not have managed to be quite so reasonable. Of course, Lavoisier knew that he was writing for the
record. As always, he took care to set the balance straight.

Notes for Further Reading

Jean-Pierre Poirier's *Antoine Laurent Lavoisier: 1743–1794* stands as the definitive Lavoisier biography, treating the whole of his life and work in full detail from start to finish and, for the most part, in chronological order. Francophone readers should be advised that the English edition of this work, *Lavoisier: Chemist, Biologist, Economist*, is significantly expanded and contains a good deal of material that the French original does not.

Arthur Donovan's *Antoine Lavoisier: Science, Administration, and Revolution* is an equally valuable work. Though not so exhaustive as the Poirier volumes, it benefits from a thematic organization as opposed to Poirier's more strictly chronological approach. Donovan organizes Lavoisier's life and work by category, and does an excellent job of situating his activities (in economics and government as well as in science) in an impressively full historical context.

Torch and Crucible: The Life and Death of Antoine Lavoisier, the 1941 biography by Sidney J. French, is certainly dated but nevertheless remains interesting, and is more accessible to a general audience than the more specialized works by Donovan and Poirier. This book, like Douglas McKie's *Antoine Lavoisier: Scientist, Economist, Social Reformer* (New York: Henry Schumann, 1952), offers many human-interest details that are not found in the more current biographies. Charles Gillispie's article on Lavoisier in *Science and Polity in France at the End of the Old Regime* (Princeton: Princeton University Press, 1980), though comparatively brief, defines an attitude toward Lavoisier and his work that persists in the most recent treatments of the chemist. Henry Guerlac's *Antoine-Laurent Lavoisier: Chemist and Revolutionary* (New York: Scribner, 1975) offers one of the most quick and efficient summaries of Lavoisier's scientific career, from a writer who is also an expert in the finest details.

Henry Guerlac's *Lavoisier—The Crucial Year: The Background and Origins of His First Experiments on Combustion in 1772* is an extraordinarily thorough treatment of that subject. It is especially useful in that it includes as appendices the drafts of most of Lavoisier's presentations (public or secret) to the Academy of Sciences during this critical period—verbatim and complete with "*ratures*." The drawback for non-Francophone readers is that throughout the text Guerlac quotes Lavoisier *only* in French.

Antoine Lavoisier—The Next Crucial Year: Or, the Sources of His Quantitative Method in Chemistry by Frederic Lawrence Holmes is, as its title implies, both an homage and a retort to Guerlac's earlier work. As declared, Holmes's target is Lavoisier's "balance sheet" system of evaluating experimental results, and while he does bring that matter to light, the most interesting thing revealed, somewhat unexpectedly (and in contrast to Lavoisier's subsequent representations), is how many more misses than hits there were in the first trials of the method. Also against expectation, Holmes shows that during this key phase of his career, Lavoisier dared push his theoretical contentions much further in advance of hard evidence than his own postulations about the standards of scientific procedure should logically have permitted him to do.

The transition from alchemy to modern chemistry is an area that seems ripe for further exploration. Trevor H. Levere's *Transforming Matter: A History of Chemistry from Alchemy to the Buckyball* gives this subject (along with many others) a swift and lucid treatment. In *Alchemy Tried in the Fire: Starkey, Boyle, and the Fate of Helmontian Chemistry*, William R. Newman and Lawrence M. Principe redefine their two seventeenth-century "chymists" as intermediates in the evolution of chemistry from alchemy—convincingly demonstrating that that evolution was rather more gradual than has heretofore been thought. Marco Beretta's *The Enlightenment of Matter: The Definition of Chemistry from Agricola to Lavoisier* is a microscopically detailed treatment of this same topic, but with a special emphasis on its iconography and above all its language; Beretta shows very persuasively how Lavoisier's reform of chemical language was at least as important as his revision of chemical theory.

Another wonderful contribution from Marco Beretta is *Imaging a Career in Science: The Iconography of Antoine Laurent Lavoisier* (Canton, MA: Science History Publications, 2001), which reproduces every known image of Lavoisier, along with Marie Lavoisier's laboratory sketches, and fully discusses their provenance and their significance in the iconographical pattern that, as much as both Lavoisiers sought to create and control it, eventually grew larger than themselves.

In March of 1994, *La revue de la Musée des Arts et Métiers* published an issue devoted to Lavoisier studies, which includes among much other interesting material a reading by Poirier of the David portrait of the Lavoisier couple, and a study by Madeleine Pinault Sørenson of Madame Lavoisier's artistic career, concentrating on her instruction by David and reproducing two previously unknown sketches from that period. *Oxygen*, a play by Roald Hoffmann and Carl Djerassi, is an interesting dramatization of the race for the discovery (and above all for the credit) among Priestley, Scheele, and Lavoisier, and accurate in its essential details. The autograph letter from Scheele to Lavoisier is reproduced as a plate in the edition of the play published by Wiley-VCH in 2001.

The instruments of Lavoisier's laboratory are on display at the Musée des Arts et Métiers in Paris. The equipment is still serviceable, and in the fall of 2003 it was used in a reenactment of several of Lavoisier's key experiments. The Archives of the Academy of Sciences in Paris contains a vast trove of Lavoisier's papers. This collection is housed in the magnificent building erected as the Collège Mazarin, where Lavoisier had his early schooling—on the left bank of the Seine, facing the Louvre where the Academy of Sciences was housed during Lavoisier's membership, a point about halfway between his Arsenal laboratory and the place of his execution.

Notes

I. Ancien régime

p. 2   "it is not from distrust . . .": Jean-Pierre Poirier, *Antoine Laurent Lavoisier: 1743–1794* (Paris: Pygmalion/Gérard Watelet, 1993), p. 335. This French edition will hereafter be noted as "Poirier, French ed."

p. 4   "I denounce to you . . .": Jean-Pierre Poirier, *Lavoisier: Chemist, Biologist, Economist*, trans. Rebecca Balinski (Philadelphia: University of Pennsylvania Press, 1996), p. 275. This translation will hereafter be noted as "Poirier."

p. 8   "Whether rectitude of the heart . . .": Sidney J. French, *Torch and Crucible: The Life and Death of Antoine Lavoisier* (Princeton: Princeton University Press, 1941), p. 20; Poirier, p. 6.

p. 8   Lacaille had taken . . .: Poirier, French ed., p. 6.

p. 9   "regulate your studies, . . .": Ibid., p. 7.

p. 10   "their active power . . .": Arthur Donovan, *Antoine Lavoisier: Science, Administration, and Revolution* (Cambridge: Cambridge University Press, 1993), pp. 34–35.

p. 10   "ingenious explication by which. . . .": Poirier, French ed., p. 15; Poirier, p. 14.

p. 12   "That's just fine! . . .": Poirier, p. 28.

p. 13   "very blue eyes, . . .": Poirier, French ed., p. 41.

p. 13   "a fool, . . .": Donovan, *Antoine Lavoisier*, p. 111.

p. 17   "In the midst of . . .": Ibid., p. 128.

p. 17   "A fluid whose movements . . .": Poirier, French ed., p. 46.

p. 18   "to finance, munitions, . . .": Donovan, *Antoine Lavoisier*, p. 125.

p. 18   "a Controller-General . . .": Poirier, p. 101.

p. 19   "without too grand . . .": Poirier, French ed., p. 49.

p. 20   "the great corps . . .": Poirier, p. 275.

p. 21   "The emoluments I enjoy . . .": Poirier, French ed., p. 291.

p. 22   "North America owes . . .": Donovan, *Antoine Lavoisier*, p. 199.

p. 23   "every exclusive privilege . . .": Ibid., p. 200.

p. 24   "It was for him . . .": Poirier, French ed., pp. 103–4.

p. 28   "*Le mur murant Paris* . . .": Poirier, p. 172.

p. 28   "If you ask what he has done . . .": Poirier, French ed., p. 210.

p. 29   "It would be useless . . .": Ibid., p. 122.

p. 30   "There is water . . .": Ibid., p. 37.

p. 32   "imagination in the absence . . .": Donovan, *Antoine Lavoisier*, p. 221.

p. 33   "The empire of the sciences . . .": Ibid., p. 227.

p. 33   "The art of drawing conclusions . . .": Poirier, French ed., p. 169.

II. Out of Alchemy

p. 34   "I was surprised to see . . .": Poirier, French ed., p. 5.

p. 35   "accustomed to the rigor . . .": Ibid., p. 6.

p. 35   "an art which teaches us . . .": Marco Beretta, *The Enlightenment of Matter: The Definition of Chemistry from Agricola to Lavoisier* (Canton, MA: Watson Publishing International, 1993), p. 131.

p. 36   "The celebrated professor . . .": Ibid., p. 132.

p. 36   "I managed to gain . . .": Donovan, *Antoine Lavoisier*, p. 47.

p. 37   "a lunatic, melancholy fantasy": William R. Newman and Lawrence M. Principe, *Alchemy Tried in the Fire: Starkey, Boyle, and the Fate of Helmontian Chemistry* (Chicago: University of Chicago Press, 2002), p. 53.

p. 37   "begins with Bernard . . .": Ibid., p. 192.

p. 38   "First he [Starkey] notes . . .": Ibid., pp. 192–93

p. 39   "the alchemist . . . worked alone. . . .": Beretta, *Enlightenment*, p. 77; C. G. Jung, *Psychology and Alchemy*, 2d ed., trans. R. F. C. Hull (Princeton: Princeton University Press, 1968), p. 179.

p. 40   "Magic has the power ...": Beretta, *Enlightenment*, p. 77.

p. 40   "In alchemy, the laboratory ...": Ibid.; Robert Halleux, "Pratique industrielle et chimie philosophique de l'Antiquité au XVII siecle," *L'Actuelité chimique* (January–February 1987), p. 19.

p. 41   "all are difficult to follow, ...": Beretta, *Enlightenment*, p. 86.

p. 45   "Within this vast and ...": Ibid., p. 109.

p. 46   "That the Fire ...": Robert Boyle, *The Skeptical Chymist, or Chymico-Physical Doubts and Paradoxes* (London: J. Cadwell, 1661), p. 73.

p. 46   "the Burning of Wood, ...": Ibid., p. 89.

p. 47   "a Mixture of Colliquated ... in Vipers": Ibid., p. 124.

p. 47   "the Fire may sometimes ...": Ibid., *Skeptical Chymist*, p. 135.

p. 48   "in terms of the interaction ...": Trevor H. Levere, *Transforming Matter: A History of Chemistry from Alchemy to the Buckyball* (Baltimore: Johns Hopkins University Press, 2001), p. 44.

p. 48   "the art of dissolving ...": Beretta, *Enlightenment*, p. 127.

p. 49   "When Galileo caused balls, ...": Ibid., p. 133; Immanuel Kant, *Critique of Pure Reason*, trans. Norman Kemp Smith (London, 1973), p. 20.

p. 50   "The philosopher's stone is ...": Poirier, French ed., p. 9.

p. 50   "For the first time ...": Beretta, *Enlightenment*, p. 129.

p. 51   "We owe to Newton ...": Antoine Nicolas Condorcet, *The Sketch for a Historical Picture of the Progress of the Human Mind*, trans. June Barraclough (Westport, CT: Greenwood Press, 1979), pp. 148–49.

p. 51   "Newton perhaps did more ...": Ibid., p. 151; see also pp. 153*ff.* for discussion of the chemical revolution.

p. 52   "experimental physics and has escaped ...": Donovan, *Antoine Lavoisier*, p. 70.

p. 53   "a highly formalized way ...": Ibid.

p. 53   "dying of chagrin from it all": Ibid., p. 62.

p. 54   "The only way to prevent errors ...": Ibid., p. 71

p. 54   "Reason must be continually ...": Ibid.

p. 57   "Nothing is created ...": Ibid., p. 181.

p. 57   "Men should frequently ...": Chemteam Web site: http://dbhs.wvusd.k12.ca.us/webdocs/Equations/Conserv-of-Mass.html. Accessed November 15, 2003.

p. 57   "Wrongly do the Greeks . . .": Ibid.

p. 58   "A fluid whose movements . . .": Poirier, p. 43.

p. 60   "If ever there was anything . . .":, Ibid., p. 63.

p. 60   "The increase in weight . . .": Ibid.

p. 61   "Before beginning the long series . . .": Poirier, French ed., p. 72.

III. Le principe oxygine

p. 64   "It has been about eight days . . .": Henry Guerlac, *Lavoisier—The Crucial Year: The Background and Origins of His First Experiments on Combustion in 1772* (Ithaca: Cornell University Press, 1961), pp. 227*ff.* Italics added.

p. 68   "was not always in perfect . . .": Ibid., p. 27.

p. 68   "Air is not a separate . . .": Poirier, p. 10.

p. 72   "all metals or metallic substance . . .": Guerlac, *Lavoisier—Crucial*, p. 209.

p. 72   "It is easy to see . . .": Ibid., p. 208.

p. 72   "the fire which chemists . . .": Ibid., p. 210.

p. 73   "It seems Constant . . .": Ibid., p. 214.

p. 74   "We will not follow . . . These views if followed . . .": Ibid.

p. 75   "I beg the public . . .": Ibid., p. 223.

p. 76   "Matter of fire . . .": Ibid., p. 219.

p. 76   "What we have just said . . .": Ibid., p. 216.

p. 77   "fire enters into . . .": Ibid. p. 22.

p. 77   "But does air exist . . .": Ibid. pp. 222–23.

p. 77   "we don't have any new . . .": Ibid., p. 215.

p. 78   "All metals exposed to fire . . .": Ibid., p. 144.

p. 79   "a machine capable . . .": Donovan, *Antoine Lavoisier*, p. 97.

p. 80   "It Would be much . . .": Guerlac, *Lavoisier—Crucial*, p. 214.

p. 81   "a volume at least . . .": Ibid., p. 228.

p. 83   "Completely Confirmed my Conjectures": Ibid., pp. 227*ff.*

p. 83   "one of the most interesting . . .": Ibid.

p. 83   "a revolution in physics and chemistry": Poirier, French ed., p. 72.

p. 86   "I know your exactitude . . .": Ibid., p. 58.

p. 86   "The more the facts . . .": Guerlac, *Lavoisier—Crucial*, facing p. 1.

p. 87   "This manner of viewing . . .": Frederic Lawrence Holmes, *Antoine*

Lavoisier—The Next Crucial Year: Or, the Sources of His Quantitative Method in Chemistry (Princeton: Princeton University Press, 1998), p. 16.

p. 88   "machine to test the effect . . .": Ibid., p. 21.

p. 89   "If [a big 'if,' under the experimental circumstances] instead of doing these experiments . . .": Ibid. p. 164.

p. 90   "It obviously results . . .": Ibid. p. 165.

p. 90   "This theory is destructive . . .": Ibid.

p. 90   "I have even come to the point . . .": Ibid., p. 37.

p. 91   "I should like to hear . . .": Ibid., p. 81.

p. 94   "all of the air . . .": Ibid., 133. Italics added.

p. 95   "submitted all his results . . .": Ibid., p. 135.

p. 95   "M. Lavoisier, so far from . . .": Ibid.

p. 98   "A candle burned . . .": Poirier, p. 74.

p. 98   "of the genuineness . . .": Ibid., p. 75.

p. 98   "saying that it was . . .": Ibid.

p. 99   "I fancied that my breast . . .": Ibid., p. 76.

p. 101 "*Monsieur, I have received* . . .": Archives of the Academy of Sciences, Paris, Lavoisier Collection, dation Chabrol, carton 2, file 3.

p. 103 "My friend, you know . . .": Poirier, p. 81.

p. 103 "We twice tried . . .": Poirier, French ed., p. 85.

p. 104 "eminently breathable air": Poirier, p. 103.

p. 104 "the principle which unites itself . . .": Poirier, French ed., p. 85.

p. 105 "After I left Paris, . . .": Poirier, p. 80.

p. 106 "a fabric woven . . .": Donovan, *Antoine Lavoisier*, p. 139.

p. 106 "a part of the experiments . . .": Poirier, French ed., pp. 116–17.

p. 107 "Here," Lavoisier declared, "is the most complete . . .": Ibid., p. 114.

p. 108 "I shall henceforward designate . . ." Beretta, *Enlightenment*, p. 177.

p. 111 "I will, of course, be asked . . .": Donovan, *Antoine Lavoisier*, p. 151.

p. 112 "without supposing that there . . .": Ibid.

p. 112 "then the Stahlian system . . .": Ibid.

p. 112 "inevitably fall into . . .": Ibid., p. 150.

p. 113 "inflammable air [hydrogen] is . . .": Poirier, p. 142.

p. 114 "if one burns a little less . . .": Ibid., p. 143.

p. 116 "It is a noble machine. . . .": Ibid., p. 151.

p. 117 "water is not an element, . . .": Ibid.

IV. The Chemical Revolution

p. 121   "Those who have seen it . . .": Madeleine Pinault Sørenson, "Madame Lavoisier: Dessinatrice et Peintre," *La revue de la Musée des Arts et Métiers* (Paris, March 1994), p. 25.

p. 122   "Madame appears . . . had been unlucky.": Poirier, p. 245.

p. 126   "This entity, introduced . . .": Beretta, *Enlightenment*, p. 181.

p. 126   "Chemists have made of phologiston . . .": Poirier, French ed., p. 191.

p. 127   "Dangerous though the spirit of . . .": Beretta, *Enlightenment*, p. 178.

p. 127   "to show that Stahl's phlogiston . . .": Ibid., p. 183.

p. 128   "I see with much satisfaction . . .": Ibid.

p. 130   Lavoisier's "antiphlogistic hypothesis" was "recommendable . . .": Poirier, p. 178.

p. 132   "Languages don't have . . .": Poirier, French ed., p. 197.

p. 132   "The art of speaking . . .": Ibid., p. 198.

p. 133   "The word ought to bring . . .": Poirier, p. 187.

p. 133   "the language of algebra . . .": Beretta, *Enlightenment*, p. 200.

p. 133   "To analyse, then, . . .": Ibid., p. 195.

p. 133   "which is necessary . . . it has marked.": Poirier, French ed., p. 197.

p. 134   "We are very far . . .": Beretta, *Enlightenment*, p. 202.

p. 134   "we are not obliged . . .": Ibid., p. 209.

p. 135   "I suppose that I have to analyze . . .": Ibid., p. 220.

p. 136   "must either reject . . .": Ibid., p. 201.

p. 137   "there is nothing else . . .": Ibid., p. 222.

p. 137   "These names have evidently . . .": Ibid., pp. 232–23.

p. 138   "that *itch* . . . for system.": Ibid., p. 233.

p. 138   "the impropriety of systematic . . .": Ibid., p. 224.

p. 138   "it is premature, . . .": Ibid., p. 235.

p. 138   "It must be a very useful . . .": Ibid., p. 236.

p. 138   "want the revolution . . .": Ibid.

p. 139   "*a method of naming* . . .": Ibid., p. 239.

p. 139   "The study of chymistry . . .": Ibid., p. 244.

p. 140   "It is by no means . . .": Antoine Lavoisier, *Elements of Chemistry* (New York: Dover Publications, 1965), pp. xvii–xviii.

p. 141   "Thoroughly convinced of these truths, . . .": Ibid., pp. xviii–xix.

p. 141   "It is not to the history . . .": Ibid., p. xxxiii.

p. 142   "that I have stated . . .": Ibid., p. xxxii.

p. 142   "long dissertations . . . to beginners": Ibid., p. xxxiii.

p. 142   "Instead of applying observation . . .": Ibid., p. xxxvi.

p. 143   "when errors have been thus . . .": Ibid.

p. 143   "But, after all, the sciences . . .": Ibid., p. xxxviii.

p. 144   "We may lay it down . . .": Ibid. p. 130.

p. 144   "All that can be said . . .": Ibid., p. xxiv.

p. 146   "swift and necessary . . . order of ideas.": Poirier, French ed., p. 197.

p. 147   *"Sir and most illustrious colleague: . . . accomplished irretrievably.*: Beretta, *Enlightenment*, p. 251.

V. The End of the Year One

p. 152   "renounce all financial . . . brothers and friends.": Poirier, p. 225.

p. 153   "extreme prudence, . . . carried against him.": Poirier, French ed., p. 240.

p. 159   "The wild *gas*, . . .": Levere, *Transforming Matter*, p. 57.

p. 160   "We regard [the political revolution] as complete . . .": Poirier, French ed. , p. 271.

p. 160   "At this moment we regret . . .": Ibid.

p. 161   "General Farmer and . . .": Ibid., p. 292.

p. 162   *"Généreux Lavoisier, ta lettre* . . .": Ibid., p. 291.

p. 163   "there I have established myself . . .": Ibid., p. 293.

p. 167   "Let them be closed . . .": Ibid., p. 355.

p. 168   "the great mass of knowledge . . . persecution of savants.": Ibid.

p. 168   "Reason was . . .": Charles Murray, "Well, It Seemed Like a Good Idea at the Time," Week in Review, *New York Times*, November 30, 2003.

p. 169   "Newton worshipers . . .": Ibid.

p. 169   "allowed for an intractable . . .": Ibid.

p. 170   "The sole foundation . . .": Condorcet, *Sketch for a Historical Picture*, p. 173.

p. 170   "The time will therefore come . . .": Ibid., p. 179.

p. 171   "the Supreme Being, . . .": George Rudé, ed., *Robespierre* (New York: Prentice Hall, 1967), p. 73.

p. 172   "The principal concern . . . owes only death.": Ibid., p. 59.

p. 175   "does honor to . . .": Poirier, French ed., p. 372.

p. 176 "It is not with celebrations . . .": Poirier, p. 340.

p. 176 "*Lavoisier, of the former . . . accounts.*": Poirier, French ed., p. 375.

p. 177 "*In a state of doubt, . . .*": Ibid.

p. 177 "*Lavoisier, of the former . . . to obey.*": Ibid.

p. 179 "I foresaw that . . .": Poirier, pp. 353.

p. 179 "minting coins on . . .": Ibid., p. 368

p. 180 "they had a technocratic . . .": Ibid., p. 363.

p. 181 "If a man is so cowardly . . .": Ibid., p. 384.

p. 182 "He spoke in favor . . .": Poirier, French ed., p. 411.

p. 183 "I don't cling to life . . .": Poirier, p. 398.

p. 183 "What galls me . . .": Poirier, French ed., p. 407.

p. 184 "prepared the others . . .": Poirier, p. 381.

p. 184 "It took them no more . . .": Ibid., p. 409.

p. 184 "You give yourself, . . .": Poirier, French ed., p. 357.

p. 185 "*I have had a reasonably long . . .*": Ibid., p. 401.

Illustration Credits

Page 15: Private Collection/Archives Charmet/Bridgeman Art Library
Page 61: Archives Charmet, Paris/Bridgemen Art Library
Page 82: Rare Book & Special Collections, University of Sydney Library
Page 99: National Library of Medicine
Page 118: National Library of Medicine
Page 124: Private Collection/Archives Charmet/Bridgemen Art Library

Index

Page numbers in *italics* refer to illustrations.